GOODS OF THE MIND, LLC

Competitive Mathematics

for

Gifted Students - Level 1 Combo

PRACTICE COUNTING

PRACTICE OBSERVATION AND LOGIC

PRACTICE ARITHMETIC

PRACTICE OPERATIONS

Cleo Borac, M. Sc.
Silviu Borac, Ph. D.

Send all inquiries to:

Goods of the Mind, LLC
1138 Grand Teton Dr.
Pacifica
CA, 94044

Competitive Mathematics for Gifted Students - Level 1 Combo
Level I (Grades 1 and 2)

Contents

FOREWORD

The goal of these booklets is to provide a problem solving training ground starting from the earliest years of a student's mathematical development.

In our experience, we have found that teaching how to solve problems should focus not only on finding correct answers but also on finding better solution strategies. While the correct answer to a problem can typically be obtained in several different ways, not all these ways are equally useful for learning how to solve problems.

The most basic strategy is *brute force*. For example, if a problem asks for the number of ways Lila and Dina can sit on a bench, it is easy to write down all the possibilities: Dina, Lila and Lila, Dina. We arrive at this solution by performing all the possible actions allowed by the problem, leaving nothing to the imagination. For this last reason, this approach is called brute force.

Obviously, if we had to figure out the number of ways 30 people could stand in a line, then brute force would not be as practical, as it would take a prohibitively long time to apply.

Using brute force to obtain the correct answer for a simpler problem is not necessarily a useful learning experience for solving a similar problem that is more complex. Moreover, solving problems in a quantitative manner, assuming that the student can transfer simple strategies to similar but more complex problems, is not an efficient way of learning problem solving.

From this simple example, we see that the goal of *practicing* problem solving is different from the goal of problem solving. While the goal of problem solving is to obtain a correct answer, the goal of practicing problem solving is to acquire the ability to develop strategies, generate ideas, and combine approaches that are powerful enough to solve the problem at hand as well as future similar problems.

While brute force is not a useless strategy, it is not a key that opens every door. Nevertheless, there are problems where brute force can be a useful tool. For instance, brute force can be used as a first step in solving a complex problem: a smaller scale example can be approached using brute force to help the problem solver understand the mechanics of the problem and generate ideas for solving the larger case.

All too often, we encounter students who can quickly solve simple problems by applying brute force and who become frustrated when the solving methods they have been employing successfully for years become inefficient once problems increase in complexity. Often, neither the student nor the parent has a clear understanding of why the student has stagnated at a certain level. When the only arrows in the quiver are guess-and-check and brute force, the ability to take down larger game is limited.

Our series of books aims to address this tendency to continue on the beaten path - which usually generates so much praise for the gifted student in the early years of schooling - by offering a challenging set of questions meant to build up an understanding of the problem solving process. Solving problems should never be easy! To be useful, to represent actual training, problem solving should be challenging. There should always be a sense of difficulty, otherwise there is no elation upon finding the solution.

Indeed, practicing problem solving is important and useful only as a means of learning how to develop better strategies. We must constantly learn and invent new strategies while questioning the limitations of the strategies we are using. Obtaining the correct answer is only the natural outcome of having applied a strategy that worked for a particular problem in the time available to solve it. Obtaining the wrong answer is not necessarily a bad outcome; it provides insight into the fallacies of the method used or into the errors of execution that may have occured. As long as students manifest an interest in figuring out strategies, the process of problem solving should be rewarding in itself.

Sitting and thinking in a focused manner is difficult to train, particularly since the modern lifestyle is not conducive to adopting open-ended activities. This is why we would like to encourage parents to pull back from a quantitative approach to mathematical education based on repetition, number of completed pages, and the number of correct answers. Instead, open up the time boundaries that are dedicated to math, adopt math as a game played in the family, initiate a math dialogue, and let the student take his or her time to think up clever solutions.

Figuring out strategies is much more of a game than the mechanical repetition of stepwise problem solving recipes that textbooks so profusely provide, in order to "make math easy." Mathematics is not meant to be easy; it is meant to be interesting.

Solving a problem in different ways is a good way of comparing the merits of each method - another reason for not making the correct answer the primary goal of the activity. Which method is more labor intensive, takes more time or is more prone to execution errors? These are questions that must be part of the problem solving process.

In the end, it is not the quantity of problems solved, the level of theory absorbed, or the number of solutions offered in ready-made form by so many courses and camps, but the willingness to ask questions, understand and explore limitations, and derive new information from scratch, that are the cornerstones of a sound training for problem solvers.

These booklets are not a complete guide to the problem solving universe, but they are meant to help parents and educators work in the direction that, aside from being the most efficient, is the more interesting and rewarding one.

The series is designed for mathematically gifted students. Each book addresses an age range as some students will be ready for this content earlier, others later. If a topic seems too difficult, simply try it again in a couple of months.

PRACTICE COUNTING

COUNTING LINEAR PATTERNS

Linear patterns are *strings* of information. In the following, we work on counting objects that are arranged in a string.

Imagine making a string of balls and rods out of magnets. The rods and balls alternate, like this:

etc.

Experiment

1. Make or draw a string of any length you want but make sure it has rods at both ends.
2. Count the number of rods and the number of balls.
3. Did you notice there is one more rod than there are balls?

Experiment

Match each pattern to the condition it satisfies:

1. ○ □ ○ □ ○ □ etc. • • •

2. ○ ○ ○ □ □ ○ □ ○ ○ □ ○ ○

3. □ ○ ○ □ ○ ○ etc. • • •

4. ○ ○ ○ □ □ □ ○ ○ ○ □ □ □ etc. • • •

(A) an apparently random sequence of squares and circles

(B) a sequence in which there are never only two circles between any two squares

(C) a sequence of alternating squares and circles

(D) a sequence in which there is always an even number of circles between any two squares

Answers: 1C, 2A, 4B, 3D

PRACTICE ONE

Exercise 1

If you have an alternating string of rods and balls with rods at both ends and 20 balls in total, how many rods does it contain?

Exercise 2

Suppose you have an alternating string of rods and balls with a total of 25 pieces. There is a ball at the left end. How many rods does it contain?

Exercise 3

If the rods and balls are really small and Dina makes a string using 1000 alternating rods and balls, could it have balls at both ends?

Exercise 4

Imagine you have a string of 513 alternating balls and rods. There is a rod at the left end, so there must be a .... at the right end.

Exercise 5

If an alternating string has balls at both ends and 31 rods, how many balls does it contain?

Exercise 6

If an alternating string is made up of 193 pieces and an odd number of rods, is the number of balls even or odd?

Exercise 7

If an alternating string is made up of 200 pieces, is the number of rods even or odd?

Exercise 8

If an alternating string is made up of 202 pieces, is the number of rods even or odd?

Exercise 9

Dina's mother makes fruit smoothie for breakfast every second day. On the other days, she makes chocolate milk. If Dina has milk this Monday, how many days will pass until she has a fruit smoothie on a Wednesday?

Exercise 10

Lila has 16 beads, some white and some red, that she wants to string together. After counting the number of beads of each color, she realizes she could make a string using a pattern of one white bead followed by two red beads. If she starts with a white bead, what color must the last bead on the string be?

Exercise 11

Arbax, the Dalmatian, has some large bones and some small bones in his bone collection. He wants to use these bones to leave messages for his friend Lynda, the terrier who lives across the street. Since Lynda only just started to learn how to read bone messages, Arbax wants to use at most two bones for each message. For example, if he wanted to tell Lynda to come over for dinner, he might place a small bone at the door of her house. Arbax cannot know if Lynda will find the message when coming home or when going out, so he cannot use "small bone, big bone" and "big bone, small bone" as two different messages. How many of the following messages can Arbax leave Lynda using his system?

   (a) Let's go for a walk!
   (b) I found a fish. Let's roll in it!
   (c) I'm going to take a nap.
   (d) Gopher alert. Come help!
   (e) Will be out of town this weekend.

CHAPTER 3. PRACTICE ONE

(f) I found a fresh tennis ball. Let's play!

(g) No treats today. I ate the cat's food again.

Exercise 12

Stephan wants to practice endurance swimming. After he swims 123 lengths of the pool, is he at the end where he left his flip-flops when he got in or at the other end?

Exercise 13

Dina is threading some red, blue, and green beads. She follows these rules: a red bead is always followed by a green bead, a green bead may be followed by a red bead or by a blue bead, and a blue bead is always followed by a red bead. How many different strings can she make out of a total of 5 beads if she starts with a red bead?

Exercise 14

Arbax has built a path out of dog treats from his house to Lynda's house. Each dog-shaped treat is followed by two bone-shaped treats. There are 23 treats in total. How many of them are dog-shaped?

COUNTING CIRCULAR PATTERNS

Imagine tying up the string of balls and rods to form a necklace. The rods and balls alternate, like this:

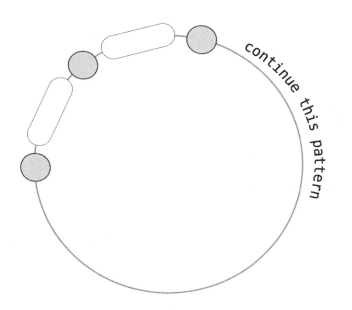

continue this pattern

Experiment

1. Make or draw a necklace of any size you want, making sure the rods and balls alternate.

2. Count the number of rods and the number of balls.

3. Did you notice that the number of balls is the same as the number of rods?

PRACTICE TWO

Exercise 1

Lila wants to make a necklace of alternating balls and rods. Will the total number of components be even or odd? Can it be either?

Exercise 2

Dina has a necklace of alternating balls and rods that has 200 components. How many rods are there?

Exercise 3

Lila has a necklace of alternating balls and rods that has 120 components. She decides the rods do not look good so she removes them. How many balls are left on the necklace?

Exercise 4

Dina has an alternating string with 101 components and balls at both ends. From one end, she wants to separate a smaller string that she could use to make a bracelet of alternating rods and balls. If the bracelet must have 20 balls, how many rods will be left on the string?

Exercise 5

Dina has a necklace with 20 pieces, all alternating balls and rods. She wants to reuse all the pieces to make two smaller alternating bracelets that are both the same size. Is this possible?

Exercise 6

Lila has a necklace with 26 components, all alternating balls and rods. She wants to reuse all the pieces to make two smaller alternating bracelets that are both the same size. Is this possible?

Exercise 7

There are 4 boys and 4 girls holding hands in a circle. How many hands are being held?

Exercise 8

Lila, Dina and three of their friends are in the garden forming a circle. Arbax scores a point if he runs from one child to the next one and gives the paw at the destination. If Arbax managed to score 15 points, how many times did he run around the circle? (Arbax runs clockwise.)

Exercise 9

Lila, Dina and three of their friends are in the garden forming a circle. Arbax has to run from child to child forming triangles. Arbax must always start and end a triangle at Dina's location. If Arbax gets a point for each different triangle, what is the largest number of points that Arbax can score?

Exercise 10

Amira thought it would be a good joke to tie Daddy's shoes and boots together to form a circle. If Amira knots the laces of neighboring shoes, how many knots must she make if Daddy has 6 pairs of shoes and boots?

Exercise 11

Arbax is holding a meeting with all the dogs on the block. The dogs are standing in a circle, with equal distances between neighbors. Arbax is directly opposite to Lynda. Which of the following statements are true?

(**A**) There is an even number of dogs.

(**B**) There is an odd number of dogs.

(**C**) We cannot say whether the number of dogs is even or odd.

(**D**) If we omit the words "with equal distances between neighbors," then (C) is true.

Exercise 12

Amira has a bracelet made of 4 orange-colored stones. At least how many grey stones must she add to the bracelet so that each grey stone has at least one grey neighbor?

Exercise 13

Which one of the following bracelets is different from the others? If needed, make a bracelet out of beads and experiment.

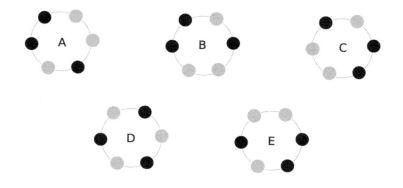

Exercise 14

Dina has a string of beads. She wants to make a bracelet out of it without breaking it. If Dina wants the bracelet to look like the one in the figure, at least how many white beads from the string should be painted grey and at least how many grey beads should be painted white?

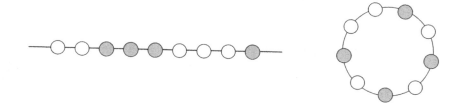

COUNTING SHARED PATTERNS

Imagine using matches to make a pattern made up of polygons, like in the following examples:

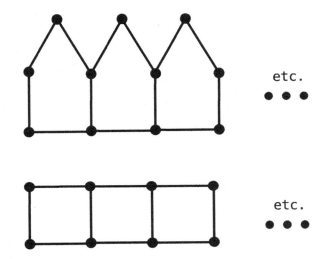

etc.
• • •

etc.
• • •

What would be a good strategy to count the total number of matches used?

Counting the matches is tricky because some of the matches are **shared** by two neighboring (we call them *adjacent*) polygons. Take a look at the following example:

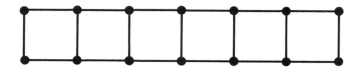

This pattern is made up of 6 squares. We notice that each individual square is made up of 4 matches, but simply multiplying 6 and 4 together will give us an incorrect answer of 24. We can easily count the matches one by one to find the correct answer: but is there a better strategy?

A good strategy for counting them is to count the sides of a basic polygon, such as this one:

and the number of additional sides needed to repeat it in an adjacent position:

Then, we multiply the number of repetitions by the number of additional sides and add the result to the number of sides of the basic figure.

The total number of matches can be computed as follows:

1. The number of matches in the basic figure is 4.
2. The number of matches in the additional figure is 3.
3. The number of additional figures is 5.
4. The total number of matches is $4 + 3 \times 5 = 19$.

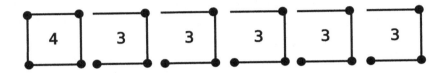

Example 1

How many sticks are needed to build the following row of 5 hexagons?

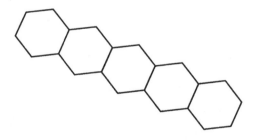

A basic figure needs 6 sticks. Each additional repetition adds 5 new sticks, for a total of 26 sticks:

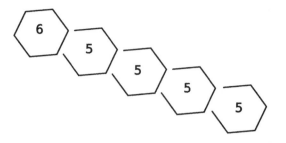

$$6 + 4 \times 5 = 26$$

Example 2

How many sticks are needed to build the following row of 50 hexagons?

Note that, even though only 4 hexagons are visible in the picture, the meaning of the dots is "We don't want to draw all 50 hexagons. After all, they are all the same. We are going to draw a few of them, to give you an idea of how the pattern goes, and then we are going to replace the rest of them with these dots."

26

The total number of sticks is:

$$6 + 49 \times 5 = 6 + 245 = 251$$

Example 3

How many sticks are needed to build a row of 500 hexagons?

The total number of sticks is:

$$6 + 499 \times 5 = 6 + 1495 = 1501$$

Experiment

Imagine more complicated patterns that share edges and apply the same counting method.

PRACTICE THREE

Exercise 1

Hannah writes her name repeatedly, but without re-writing the letter "H," like this:

HANNAHANNAHAN...

Which is the 43rd letter she writes?

Exercise 2

Write all the numbers from 6 to 15 in a string, without any spaces. What is the middle digit?

Exercise 3

Write all the numbers from 6 to 95 in a string, without any spaces. What is the middle digit?

Exercise 4

If we have 6 separate squares made of matches, how many matches must we remove if we want to make a row of squares where neighboring squares share one side?

Exercise 5

If we have 600 separate squares made of matches, how many matches must we remove if we want to make a row of squares where neighboring squares share one side?

Exercise 6

A row of squares in which neighboring squares share a side has been made out of 13 matches. How many squares are there in the row?

Exercise 7

What is the smallest number of sticks of any length that could be used to build the following figure?

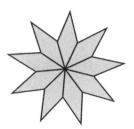

Exercise 8

What is the smallest number of sticks of any length that could be used to build the following figure?

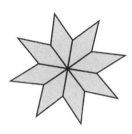

Exercise 9

A 4 × 4 checkerboard has:

(A) the same number of white squares as black squares

(B) more white squares than black squares

(C) fewer white squares than black squares

(D) whether there are more white squares is something that depends on the color of the corners

Exercise 10

A 5×5 checkerboard has:

(A) the same number of white squares as black squares

(B) more white squares than black squares

(C) fewer white squares than black squares

(D) whether there are more white squares is something that depends on the color of the corners

Exercise 11

A 42×42 checkerboard has:

(A) the same number of white squares as black squares

(B) more white squares than black squares

(C) fewer white squares than black squares

(D) whether there are more white squares is something that depends on the color of the corners

Exercise 12

A 51×51 checkerboard has:

(A) the same number of white squares as black squares

(B) more white squares than black squares

(C) fewer white squares than black squares

(D) whether there are more white squares is something that depends on the color of the corners

Exercise 13

Lila wants to build the grid in the figure out of matches. Each small side is made out of one match. How many matches will she need?

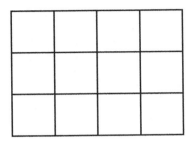

Exercise 14

Lila wants to build the grid in the figure out of matches. Each small side is made out of one match. How many matches will she need?

24 small squares in total

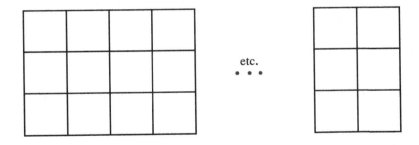

etc.

Exercise 15

Dina has 8 sticks of equal lengths. What is the largest number of squares of any size she can build using them?

TABLES

Tables are formed of *cells*. Cells are arranged in *rows* and *columns*. This is a table with 4 rows and 5 columns:

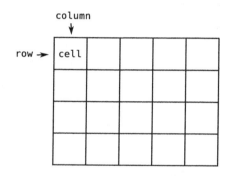

In each cell we may place an *element*. Elements can be numbers, words or other kinds of objects.

Columns are counted *from left to right*, with the leftmost column first.

Rows are counted *from top to bottom*, with the topmost row first.

The number of rows and the number of columns are the *dimensions* of the table.

We often write "a 3×4 table" if we mean a table with 3 rows and 4 columns. We also say, verbally, "a three **by** four table," to mean the same thing.

Experiment

Make a table with 3 rows and 2 columns.

Make a 2×7 table.

PRACTICE FOUR

Exercise 1

In the table below, the 3rd column has been shaded in. Shade the 2nd row as well. How many cells belong to both the 3rd column and the 2nd row?

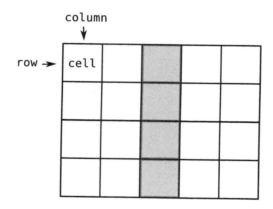

Exercise 2

How many cells are there in a 3 × 5 table?

Exercise 3

If we move on a diagonal in a 5 × 3 table, what is the largest number of cells we can visit?

Exercise 4

In the following table, is it possible to color half the cells red and half the cells blue?

Exercise 5

In the following table, color all cells that are on even rows and, at the same time, on even columns:

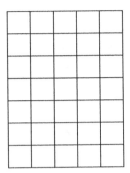

How many cells have you colored?

Exercise 6

Make a 5 × 4 table and shade all the cells that have an even row number.

Exercise 7

Make a 5 × 4 table and shade all the cells that have an odd column number.

Exercise 8

Make a 5 × 4 table and shade all the cells for which the sum of the row and column numbers is even.

Exercise 9

Make a 5 × 4 table and shade all the cells for which the sum of the row and column numbers is odd.

Exercise 10

Make a 4 × 4 table and shade all the cells for which the sum of the row and column numbers is 5.

Exercise 11

To help count by fives, you can place the counting numbers in a table with 5 rows, like this:

0	5	10	15	20	25
1	6	11	16	21	26
2	7	12	17	22	27
3	8	13	18	23	28
4	9	14	19	24	. . .

. . .

The table would have an infinite number of columns, which is why we have used some dots to show that it just keeps going on.

If you look carefully, you notice that you can count by fives if you read the first row of the table.

1. What do the other rows of the table help you do?
2. How many rows should a similar table have in order to help you count by sevens?

NUMBER PATTERNS

Whole (integer) numbers that can be arranged to form a sequence of numbers that always increase by 1 are called *consecutive numbers*. You can form a sequence of consecutive numbers by starting with any number you like and counting up by 1, as in the example below:

$$4, 5, 6, 7, 8, 9, 10, 11$$

Consecutive even numbers are even numbers that can be arranged to form a sequence increasing by 2s:

$$12, 14, 16, 18, 20, 22, 24$$

Consecutive odd numbers are odd numbers that can be arranged to form a sequence increasing by 2s:

$$1, 3, 5, 7, 9, 11$$

Experiment

Make a list of 5 consecutive numbers starting from 11.

Make a list of 6 consecutive even numbers starting from 0.

Make a list of 7 consecutive odd numbers starting from 15.

PRACTICE FIVE

Exercise 1

Are the numbers in this box consecutive even numbers?

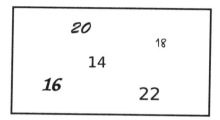

Exercise 2

We want to arrange the numbers in the box to form a sequence of consecutive numbers. Which numbers are we missing?

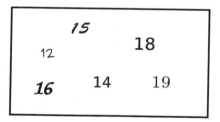

Exercise 3

We count by 2s from 2 to 20 and then we count by 2s downwards from 19 to 1. How many numbers have we counted in total?

Exercise 4

We count by 2 from 2 to 200, then we count by 2 downwards from 199 to 1. How many numbers have we counted in total?

Exercise 5

A *palindrome* is a number that remains the same if the digits are reversed. The numbers 232, 4994, and 1004001 are palindromes. List four consecutive palindromes starting from 989.

Exercise 6

Continue the following sequence of figures until there are 6 squares in a row for the first time. How many stars have been used so far?

Exercise 7

Continue the following sequence of figures until there are 5 stars in a row for the first time. How many squares have been used so far?

Exercise 8

How many different rows of figures can you make using 2 stars and 1 square in each row?

Exercise 9

You have two stars and a square. How many different rows of figures can you make using all the shapes, if the stars must remain next to each other? Draw them all.

Exercise 10

You have a red star, a blue star, and a square. How many different rows can you make using all the shapes, if the stars have to be side by side? Draw them all. How is this problem different from the previous one?

Exercise 11

Lila counted by 5s from 10 to 50 while Dina counted backwards by 10s from 50 to 0. When they were finished, Lila and Dina compared notes to see if any of the numbers they counted were the same. How many such numbers did they find?

Exercise 12

Lila counted the even numbers from 2 to 10. After each number Lila counted, Dina said an odd number that differed from Lila's number by one. The number of times Dina said "7" cannot be:

(A) 0

(B) 1

(C) 2

(D) 3

Exercise 13

The March Hare has a carrot, a red pepper, and a lettuce which he plans to eat one at a time. In how many different orders can he eat them?

Exercise 14

You have 3 potatoes and 2 beets. In how many ways can you arrange them in a row so that the row looks the same from left to right as from right to left? Draw them all.

Exercise 15

From the integers $1, 2, 3, 4, 5, 6, 7, 8, 9$, and 10, select a smaller group of integers such that no integer in the group is the double of another integer in the group. What is the largest number of integers that can form such a group?

Exercise 16

Dina said the numbers $1, 15, 9, 16$, and 23 while Lila said the numbers $11, 17, 8, 5$, and 34. How many of these numbers are larger than the smallest number Lila said and smaller than the largest number Dina said?

Exercise 17

Dina writes in increasing order a list of consecutive even numbers starting at 4. Lila writes in decreasing order a list of consecutive odd numbers starting at 19. How many numbers are there in the largest possible list of consecutive integers that can be made with numbers from both lists?

Exercise 18

Amira counted how many 2-digit numbers have an even digit in the tens place and an odd digit in the units place. Lila counted how many 2-digit numbers have an odd digit in the tens place and an even digit in the units place. Who counted more numbers?

(A) Amira

(B) Lila

(C) They both counted the same number of numbers.

MISCELLANEOUS PRACTICE

Exercise 1

Tom alternates between eating cereal and eating muffins at breakfast. If he wants to end the month eating the same breakfast as on the first day of the month, then the month can be:

(A) April

(B) June

(C) July

(D) November

(E) any month

Exercise 2

If a book has 30 pages, how many different places are there to put a bookmark in it? Assume the book has covers that are not included in the page count and begins at page one.

Exercise 3

In the *Dance of the Lilies* there are dancers dressed as pink lilies and dancers dressed as blue lilies. The blue lilies come on stage in rows of 4 dancers. The pink lilies come on stage and position themselves so that each of them has 4 blue lilies as neighbors, like this:

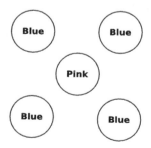

How many pink lilies are there onstage, if there are 3 rows of blue lilies?

Exercise 4

There are five chickens in a coop. Each chicken lays an egg every other day.

1. What is the largest number of eggs the chickens can produce in one week?

2. What is the largest number of eggs the chickens can produce in two weeks?

Exercise 5

Max, the baker, is counting the number of people who go in and out of his shop. Starting at 10:00 am, 3 people enter and 2 people exit the shop every 2 minutes. How many more people are there in the shop at 10:10 am than at 10:00 am?

(A) 4

(B) 5

(C) 9

(D) 10

(E) 20

Exercise 6

On the surface of a pond, there is a circle formed of waterlilies. On each lily there is a frog. Between any two lilies there is a frog. In total, there are 26 frogs. How many waterlilies are there?

Exercise 7

On a street, houses have even numbers on one side and odd numbers on the other. At one end of the street, house number 100 is across the road from house number 101. House number 202 is at the other end of the street. What is the number of the house across the road from it? (Assume all houses have the same shape and size.)

Exercise 8

Take a sheet of 8-1/2 by 11 inch letter paper and draw lines parallel to the shorter side every inch. Then, start folding along the lines in accordion style. Place the accordion on a table and notice how some creases touch the table as valleys and some point upward as hills. If the accordion has a valley at one end, does it have a valley or a hill at the other end?

Exercise 9

Cubes are used to make a square enclosure, like this:

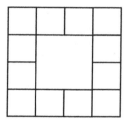

How many cubes are needed to build an enclosure that has 8 cubes on each side?

Exercise 10

A cube has 6 square faces. If you place it on a table, only 5 square faces can be seen (you are allowed to walk around it). If you make a tower of 3 cubes, how many square faces are going to be visible?

Exercise 11

A cube has 6 square faces. If you place it on a table, only 5 square faces can be seen (you are allowed to walk around it). If you make a tower of 10 cubes, how many square faces are going to be visible?

Exercise 12

How many times do I have to put my right foot in front of my left foot if I take 10 steps forward?

Exercise 13

Dina has 2 six sided dice with numbers from 1 to 6 on them. She throws them and adds the numbers that come up. How many possible results can she get?

Exercise 14

On a six sided die, we paint two faces black and the remaining faces red. How many different dice can we make?

Exercise 15

Dina has 4 square pieces of paper which she can place on a grid to make a shape. The squares must share complete edges and must not overlap. The squares have the same color on both sides. How many different shapes can Dina make?

Exercise 16

Lila is making an abacus of her own design. On the first rod she threads one red bead and 11 blue beads, on the second rod she threads two red beads and 10 blue beads, and follows this pattern until the colors are completely reversed. How many rods does Lila's abacus have?

SOLUTIONS TO PRACTICE ONE

Exercise 1

If you have an alternating string of rods and balls with rods at both ends and 20 balls in total, how many rods does it contain?

Solution 1

If the ends are both rods, then there is one more rod than there are balls. There are 21 rods.

Exercise 2

Suppose you have an alternating string of rods and balls with a total of 25 pieces. There is a ball at the left end. How many rods does it contain?

Solution 2

Notice that:

1. If the total number of alternating objects is even, then the objects at the ends are different.

2. If the total number of alternating objects is odd, then the ends are both of the type of object that there is one more of.

Since the number of objects is odd, both ends must be balls. Therefore, the number of balls is one more than the number of rods. There are 12 rods and 13 balls.

Exercise 3

If the rods and balls are really small and Dina makes a string using 1000 alternating rods and balls, could it have balls at both ends?

Solution 3

No. Since the total number of alternating objects is even, the ends of the string must be different.

Exercise 4

Imagine you have a string of 513 alternating balls and rods. There is a rod at the left end, so there must be a .... at the right end.

Solution 4

Since the total number of alternating objects is odd, the objects at the ends of the string must be identical. The right end must be a rod as well.

Exercise 5

If an alternating string has balls at both ends and has 31 rods, how many balls does it contain?

Solution 5

The number of balls must exceed the number of rods by 1. There are 32 balls.

Exercise 6

If an alternating string is made up of 193 pieces and has an odd number of rods, is the number of balls even or odd?

Solution 6

If the total number of objects is odd then the number of rods differs from the number of balls by 1. The number of balls must be even. Alternately, use the fact that even+odd=odd.
(Notice, no computation is needed for either solution.)

Exercise 7

If an alternating string is made up of 200 pieces, is the number of rods even or odd?

Solution 7

If the total number of objects is even, then there is an equal number of objects of each type. The number of rods is exactly half the total number of pieces: 100. The number of rods must be even.

Exercise 8

If an alternating string is made up of 202 pieces, is the number of rods even or odd?

Solution 8

If the total number of objects is even, then there are equal numbers of objects of each type. Since the half of 202 is 101, which is odd, the number of rods is odd.

Exercise 9

Dina's mother makes fruit smoothie for breakfast every second day. On the other days, she makes chocolate milk. If Dina has milk this Monday, how many days will pass until she has a fruit smoothie on a Wednesday?

Solution 9

This week, Dina's mother prepares the same drink on Monday, Wednesday, Friday, and Sunday. Since the number of days in a week is odd, consecutive weeks will have different Wednesday treats. The first week, Dina has milk on Monday and Wednesday. She will have a smoothie on the second Wednesday after the current Monday. That is, in nine days.

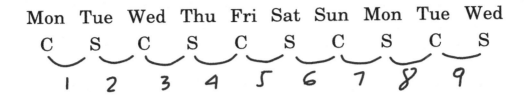

Exercise 10

Lila has 16 beads, some white and some red, that she wants to string together. After counting the number of beads of each color, she realizes she could make a string using a pattern of one white bead followed by two red beads. If she starts with a white bead, what color must the last bead on the string be?

Solution 10

The pattern is made up of groups of three beads (WRR) that follow one another: WRRWRR.... It takes Lila 15 beads to create five such groups. Since she has 16 beads in total, the last bead must be white in order to preserve the pattern.

Exercise 11

Arbax, the Dalmatian, has some large bones and some small bones in his bone collection. He wants to use these bones to leave messages for his friend Lynda, using at most two bones for each message. How many different messages can Arbax leave Lynda using his system?

Solution 11

Arbax can leave five different messages. Denote a small bone with S and a large bone with L. The possible messages are: S, L, SS, LL and SL. Since Lynda could look at the message from any direction, Arbax cannot use different messages for SL and LS.

Exercise 12

Stephan wants to practice endurance swimming. After he swims 123 lengths of the pool, is he at the end where he left his flip-flops when he got in or at the other end?

Solution 12

The pool lengths can be paired. After each pair, Stephan will be at the same end he started at. Since 123 is odd, one of the lengths cannot be paired. Stephan will get out of the pool at the opposite end from his flip-flops.

Exercise 13

Dina is threading some red, blue, and green beads. She follows these rules: a red bead is always followed by a green bead, a green bead may be followed by a red bead or by a blue bead, and a blue bead is always followed by a red bead. How many different strings can she make out of a total of 5 beads if she starts with a red bead?

Solution 13

Make a tree diagram that models the decisions made by Dina. There are only three strings that follow all the rules: RGBRG, RGRGB and, RGRGR:

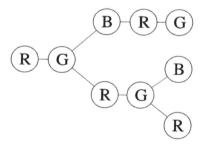

Exercise 14

Arbax has built a path out of dog treats from his house to Lynda's house. Each dog-shaped treat is followed by two bone-shaped treats. There are 23 treats in total. How many of them are dog-shaped?

50

Solution 14

The question is mainly about the kind of treats that are at the two ends. If one end is dog-shaped and the other is bone-shaped, then the total number of treats must be a number that can be formed out of groups of 3 (one dog-shaped and two bone-shaped) or of groups of 3 and one group of 2 is there is only one bone-shaped treat at one end.

$$DBB\ DBB \cdots DBB$$

or

$$DBB\ DBB \cdots DBB\ DB$$

The number 23 matches this last case since: $23 = 21 + 2 = 3 \times 7 + 2$. In this case, there are $7 + 1 = 8$ dog-shaped treats.

If we have dog-shaped treats at both ends, then the total number of treats must be a number that can be formed out of groups of 3 plus an extra treat:

$$DBB\ DBB \cdots DBB\ D$$

This is not possible if there are 23 treats in total.

If we have bone-shaped treats at both ends, then we can have:

$$B\ DBB\ DBB \cdots DBB\ DB$$

in which case the total number of treats must be formed of groups of 3 exactly (you can see this easily if you move the first B to the end), or

$$BBD\ BBD \cdots BBD\ B$$

in which case there are groups of 3 plus one, or

$$BBD\ BBD \cdots BBD\ BB$$

in which case there are groups of 3 plus 2. This last case works for the number 23 because $23 = 3 \times 7 + 2$. In this case there are 7 dog-shaped treats as well.

We see that there are two possible ways in which the treats can be placed but, in both cases, the number of dog-shaped treats is 7.

SOLUTIONS TO PRACTICE TWO

Exercise 1

Lila wants to make a necklace of alternating balls and rods. Should the total number of components be even, odd or either?

Solution 1

Since the number of rods must be equal to the number of balls, the total number of objects needed to make an alternating necklace must be even.

Exercise 2

You have a necklace of alternating balls and rods that has 200 components. How many rods are there?

Solution 2

100 rods.

Exercise 3

Lila has a necklace of alternating balls and rods that has 120 components. She decides the rods do not look good so she removes them. How many balls are left on the necklace?

Solution 3

Since the number of rods must be equal to the number of rods, there are 60 balls left on the necklace.

Exercise 4

Dina has an alternating string with 101 components and balls at both ends. From one end, she wants to separate a smaller string that she could use to make a bracelet of alternating rods and balls. If the bracelet must have 20 balls, how many rods will be left on the string?

Solution 4

Since the bracelet is circular, it has to have an equal number of balls and rods. Therefore, it must have a total of 40 pieces. The string will be left with a total of $101 - 40 = 61$ pieces.

Therefore, there are 31 balls and 30 rods left on the string.

Exercise 5

Dina has a necklace with 20 pieces, all alternating balls and rods. She wants to reuse all the pieces to make two smaller alternating bracelets that are both the same size. Is this possible?

Solution 5

Yes, it is possible since each of the smaller bracelets would have 10 components. 10 is an even number, therefore it is possible to form a loop of alternating components.

Exercise 6

Lila has a necklace with 26 components, all alternating balls and rods. She wants to reuse all the pieces to make two smaller alternating bracelets that are both the same size. Is this possible?

Solution 6

No, it is not possible since each of the smaller bracelets would have 13 components. 13 is an odd number, therefore it is not possible to form a loop of alternating components.

Exercise 7

There are 4 boys and 4 girls holding hands in a circle. How many hands are being held?

Solution 7

There are 8 children in total. Since they are in a circle, all the hands are being held. 8 children have 16 hands. Therefore, the answer is 16.

Exercise 8

Lila, Dina and three of their friends are in the garden forming a circle. Arbax scores a point if he runs from one child to the next one and gives the paw at the destination. If Arbax managed to score 15 points, how many times did he run around the circle? (Arbax runs clockwise.)

Solution 8

In the diagram, the small circles represent the 5 children. Each time he runs a full circle, Arbax collects 5 points. To collect 15 points, Arbax must run around the circle 3 times.

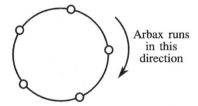

Arbax runs in this direction

Exercise 9

Lila, Dina and three of their friends are in the garden forming a circle. Arbax has to run from child to child forming triangles. Arbax must always start and end a triangle at Dina's location. If Arbax gets a point for each different triangle, what is the largest number of points that Arbax can score?

Solution 9

Arbax can make the 6 different triangles shown in the figures:

Dina   Dina

Exercise 10

Amira thought it would be a good joke to tie Daddy's shoes together to form a circle. If Amira knots the laces of neighboring shoes, how many knots must she make if Daddy has 6 pairs of shoes?

Solution 10

There are two shoes in a pair. Therefore, there are 12 shoes in total. Each shoe has one lace with two ends: one end gets tied to one neighboring shoe and the other end gets tied to the other neighboring shoe. Therefore, there is a knot between any two adjacent (neighboring) shoes. Since there are 12 shoes in a circle, there are also 12 spaces between them and, therefore, 12 knots.

Solution 11

(A) is true. If there is an even number of dogs, each dog has a dog opposite from itself on the circle. If there is an odd number of dogs, no dog has a dog opposite from itself.

However, without the information that the dogs are arranged at equal distances, it is not possible to say for sure what the parity of the number of dogs is. Some dogs may have an opposite, while others may not. (D) is also true.

Exercise 12

Amira has a bracelet made of 4 orange-colored stones. At least how many grey stones must she add to the bracelet so that each grey stone has at least one grey neighbor?

Solution 12

Amira must place at least 2 grey stones between any two neighboring orange-color stones. In total, at least 8 grey stones.

Solution 13

The correct answer is (**D**). Bracelet B can be obtained by flipping bracelet A upside-down. C is obtained by rotating A. E is obtained by rotating B.

Exercise 14

Dina has a string of beads. She wants to make a bracelet out of it without breaking it. If Dina wants the bracelet to look like the one in the figure, at least how many white beads from the string should be painted grey and at least how many grey beads should be painted white?

Solution 14

One white bead should be painted grey and one grey bead should be painted white:

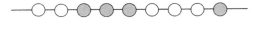

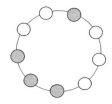

SOLUTIONS TO PRACTICE THREE

Exercise 1

Hannah writes her name repeatedly, but without re-writing the letter "H," like this:

HANNAHANNAHAN...

Which is the 43rd letter she writes?

Solution 1

This pattern is formed by repeating the following building block: HANNA

This building block has 5 letters and can be repeated twice within the span of 10 letters. It can therefore by repeated 8 times within the span of 40 letters. Three more letters remain to complete the string to 43 letters. They are the three letters at the beginning of the block: HAN. The 43rd letter is the letter N.

Exercise 2

Write all the numbers from 6 to 15 in a string, without any spaces. What is the middle digit?

Solution 2

There are 4 one-digit numbers from 6 to 9. The remaining numbers are 2-digit numbers and, regardless how many of them there are in the string, the number of digits required to write them out is even. Since 4 is even, adding an even number to it will produce an even sum. The string will have an even number of digits and, therefore, there is no middle digit.

Notice that it is not necessary to write out the string.

Exercise 3

Write all the numbers from 6 to 95 in a string, without any spaces. What is the middle digit?

Solution 3

Using the same reasoning as before, there is no middle digit.

Note to parents: Notice how, in this case, it is extremely difficult to brute force the problem, i.e. to actually write the string down. The reasoning based on parity that we derived in the previous problem scales well to larger cases and represents the correct way to solve the problem. Also, note that the solution is *non-constructive*, i.e. it only allows us to prove that there is no answer. Should there be an answer, this method would not help us construct (find) it.

Exercise 4

If we have 6 separate squares made of matches, how many matches must we remove if we want to make a row of squares where neighboring squares share one side?

Solution 4

Think of building the pattern from one basic unit and add-on units. For each add-on unit, one match is already in place. The diagram shows how to combine two squares:

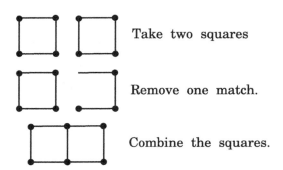

Take two squares

Remove one match.

Combine the squares.

Therefore, we must remove as many matches as there are add-on units.

For the 6 squares in a row, we have a basic unit of one square and 5 add-on units formed of only three edges. Therefore, we must remove 5 matches.

Note to parents: Note how this solution is easily scalable to a much larger number of squares, where a brute force approach (counting all the shared edges) is not practicable.

Exercise 5

If we have 600 separate squares made of matches, how many matches must we remove if we want to make a row of squares where neighboring squares share one side?

Solution 5

Think of building the pattern from one basic unit and add-on units. For each add-on unit, one match is already in place. Therefore, we must remove as many matches as there are add-on units. For the 600 squares in a row, we have a basic unit of one square and 599 add-on units formed of only three edges. Therefore, we must remove 599 matches.

Exercise 6

A row of squares in which neighboring squares share a side has been made out of 13 matches. How many squares are there in the row?

Solution 6

Imagine the row of squares broken down into one basic unit of 4 matches and a number of add-ons formed of 3 matches each. Subtract the matches used for the basic unit from the total: $13 - 4 = 9$. There are 9 matches available for add-on units. Since each add-on has 3 edges, there must be 3 add-ons. Therefore, there are 4 squares in total.

Exercise 7

What is the smallest number of sticks of any length that could be used to build the following figure?

Solution 7

Each repeating pattern is formed of 3 sticks: one interior spoke plus two outer sticks, as shown in the figure. Since there are 9 interior spokes, there are as many repetitions of the three stick pattern. The total number of sticks is 27.

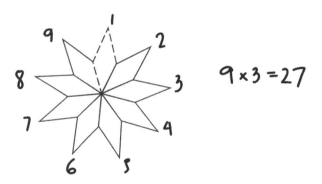

Exercise 8

What is the smallest number of sticks of any length that could be used to build the following figure?

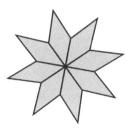

Solution 8

There are 8 interior spokes. $8 + 8 = 16$ sticks are needed to build the outline. Since the number of spokes is even only 4 sticks are needed to build the interior spokes. They will intersect at the midpoint:

The smallest number of sticks needed in total is $16 + 4 = 20$.

Solution 9

A 4×4 checkerboard has:

A. The same number of white squares as black squares.

Each row of the checkerboard has an even number of squares: half are black and half are white.

Solution 10

A 42×42 checkerboard has:

A. The same number of white squares as black squares.

Each row of the checkerboard has an even number of squares: half are black and half are white.

Solution 11

For a 5×5 checkerboard:

D. Whether there are more white squares is something that depends on the color of the corners.

Each row of the checkerboard has an odd number of squares. The number of white squares and the number of black squares differ by 1. There is an odd number of rows in total. If the first row has white corners, there are more white squares in total. Experiment with coloring 5×5 grids.

Solution 12

For a 51×51 checkerboard:

D. Whether there are more white squares is something that depends on the color of the corners.

Now you cannot experiment anymore, since the checkerboard is very large. But you can use the result from the previous problem, since the number of squares in a row is odd and the number of rows is odd as well.

Exercise 13

Lila wants to build the grid in the figure out of matches. Each small side is made out of one match. How many matches will she need?

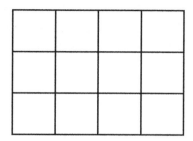

Solution 13

Find a unit that repeats:

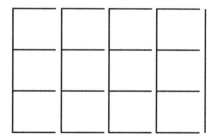

Exercise 14

Lila wants to build the grid in the figure out of matches. How many matches will she need?

Solution 14

First, notice that there must be 8 squares in each row: $8 + 8 + 8 = 24$. Then, find a pattern that repeats:

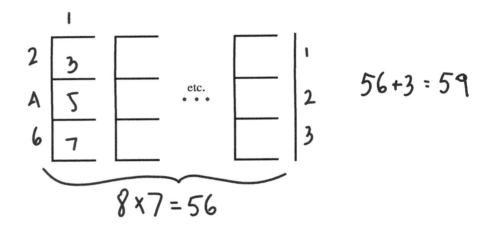

There are 7 matches in each unit that repeats. Since we need 8 units, there will be 56 matches. Finally, we have to add the 3 matches that close the figure at the rightmost end for a grand total of 59 matches.

Exercise 14

Dina has 8 sticks of equal lengths. What is the largest number of squares of any size she can build using them?

Solution 14

Dina can arrange the sticks as in the figure:

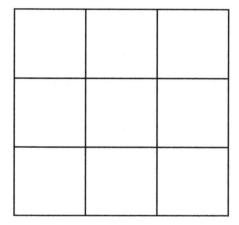

We can count 9 squares of size 1×1, 4 squares of size 2×2 and one large square of size 3×3. In total, $9 + 4 + 1 = 14$ squares of different sizes.

SOLUTIONS TO PRACTICE FOUR

Exercise 1

In the table below, the 3rd column has been shaded in. Shade the 2nd row as well. How many cells belong to both the 3rd column and the 2nd row?

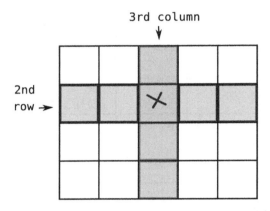

Solution 1

One cell.

Exercise 2

How many cells are there in a 3×5 table?

Solution 2

15 cells. Multiply the dimensions of the table to get the total number of cells.

Exercise 3

If we move on a diagonal in a 5×3 table, what is the largest number of cells we can visit?

Solution 3

We can visit at most 3 cells. Try to experiment with various starting cells. The following figure shows two possibilities:

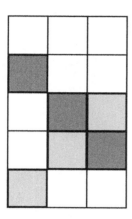

Exercise 4

In the following table, is it possible to color half the cells red and half the cells blue?

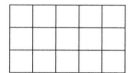

Solution 4

No, it is impossible. The table has 5 columns and 3 rows. The number of cells is therefore an odd number (15) and does not have an integer half.

Exercise 5

In the following table, color all cells that are on even rows and, at the same time, on even columns:

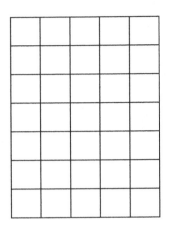

How many cells have you colored?

Solution 5

You have colored 6 cells, right?

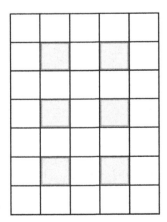

Exercise 6

Make a 5 × 4 table and shade all the cells that have an even row number.

Solution 6

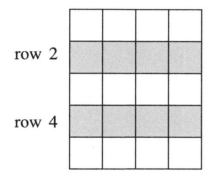

Exercise 7

Make a 5 × 4 table and shade all the cells that have an odd column number.

Solution 7

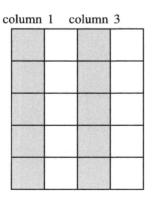

Exercise 8

Make a 5×4 table and shade all the cells for which the sum of the row and column numbers is even.

Solution 8

1+1=2		1+3=4	
	2+2=4		2+4=6
3+1=4		3+3=6	
	4+2=6		4+4=8
5+1=6		5+3=8	

Exercise 9

Make a 5×4 table and shade all the cells for which the sum of the row and column numbers is odd.

Solution 9

	1+2=3		1+4=5
2+1=3		2+3=5	
	3+2=5		3+4=7
4+1=5		4+3=7	
	5+2=7		5+4=9

Exercise 10

Make a 4×4 table and shade all the cells for which the sum of the row and column numbers is 5.

Solution 10

			1+4=5
		2+3=5	
	3+2=5		
4+1=5			

Exercise 11

To help count by fives, you can place the counting numbers in a table with 5 rows, like this:

0	5	10	15	20	25
1	6	11	16	21	26
2	7	12	17	22	27
3	8	13	18	23	28
4	9	14	19	24	· · ·

· · ·

1. What do the other rows of the table help you do?

2. How many rows should a similar table have in order to help you count by sevens?

Solution 11

1. The other rows help you: count by 5s starting from 1, count by 5s starting from 2, count by 5 starting from 3, and count by 5s starting from 4.

2. The table would have 7 rows.

SOLUTIONS TO PRACTICE FIVE

Exercise 1

Are the numbers in this box consecutive even numbers?

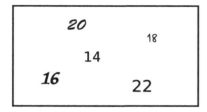

Solution 1

Yes. The numbers are: 14, 16, 18, 20, and 22.

Exercise 2

We want to arrange the numbers in the box to form a sequence of consecutive numbers. Which numbers are we missing?

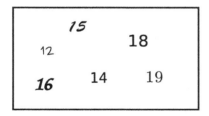

Solution 2

We are missing 13 and 17.

Exercise 3

We count by 2s from 2 to 20 and then we count by 2s downwards from

19 to 1. How many numbers have we counted in total?

Solution 3

Notice that we have counted all the numbers from 1 to 20. Therefore, we have counted 20 numbers.

Exercise 4

We count by 2 from 2 to 200, then we count by 2 downwards from 199 to 1. How many numbers have we counted in total?

Solution 4

We have counted 200 numbers in all.

Exercise 5

A *palindrome* is a number that remains the same if the digits are reversed. List four consecutive palindromes starting from 989.

Solution 5

989, 999, 1001, and 1111

Exercise 6

Continue the following sequence of figures until there are 6 squares in a row for the first time. How many stars have been used so far?

Solution 6

Notice that the odd numbers are symbolized with stars and the even numbers are symbolized with squares. The number of stars used until there are 6 squares in a row for the first time is:

$$1 + 3 + 5 = 9$$

Exercise 7

Continue the following sequence of figures until there are 5 stars in a row for the first time. How many squares have been used so far?

Solution 7

Notice that the squares separate consecutive numbers of stars. Continue the sequence with 4 stars, a square, and then 5 stars. Up to now, 4 squares have been used.

Exercise 8

How many different rows of figures can you make using 2 stars and 1 square in each row?

Solution 8

3 different rows:

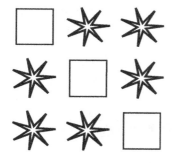

Exercise 9

You have two stars and a square. How many different rows can you make using all the shapes, if the stars have to remain next to each other? Draw them all.

Solution 9

2 different rows:

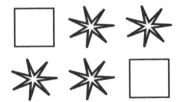

Exercise 10

You have a red star, a blue star, and a square. How many different rows of figures can you make using all the shapes, if the stars have to be side by side? Draw them all. How is this problem different from the previous one?

Solution 10

4 different rows. Use the previous problem and notice that the blue and red stars can be swapped to make two more rows:

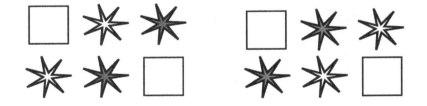

Exercise 11

Lila counted by 5s from 10 to 50 while Dina counted backwards by 10s from 50 to 0. When they were finished, Lila and Dina compared notes to see if any of the numbers they counted were the same. How many such numbers did they find?

Solution 11

5 numbers: $10, 20, 30, 40$, and 50.

Exercise 12

The correct answer is (D). Lila counted the even numbers from 2 to 10. After each number Lila counted, Dina said an odd number that differed from Lila's number by one. The number of times Dina said "7" cannot be:

(A) 0

(B) 1

(C) 2

(D) 3

Solution 12

Dina could have avoided saying 7 by saying 5 when Lila said 6 and by saying 9 when Lila said 8. Therefore, 0 is a possibility.

Dina could have said 7 only one time by saying 5 when Lila said 6 and by saying 7 when Lila said 8. Therefore, 1 is a possibility.

Dina could have said 7 twice by saying 7 when Lila said 6 and by saying 7 again when Lila said 8. Therefore, 2 is a possibility.

Since there are only two even numbers that are consecutive to 7, it is not possible for Dina to say 7 more than two times.

Exercise 13

The March Hare has a carrot, a red pepper, and a lettuce which he plans to eat one at a time. In how many different orders can he eat them?

Solution 13

Six.

- carrot, pepper, lettuce
- carrot, lettuce, pepper
- lettuce, pepper, carrot
- lettuce, carrot, pepper
- pepper, carrot, lettuce
- pepper, lettuce, carrot

Exercise 14

You have 3 potatoes and 2 beets. In how many ways can you arrange them in a row so that the row looks the same from left to right as from right to left? Draw them all.

Solution 14

In only two ways:

- potato, beet, potato, beet, potato
- beet, potato, potato, potato, beet

Exercise 15

From the integers $1, 2, 3, 4, 5, 6, 7, 8, 9$, and 10, select a smaller group of integers such that no integer in the group is the double of another integer in the group. What is the largest number of integers that can form such a group?

Solution 15

Only even numbers can be the double of an integer. The only even numbers that are not the doubles of some odd number are 4 and 8. Since 8 is the double of 4, we cannot keep them both. The strategy is to make a list of which numbers cannot be in the smaller group:

out of $\{1, 2, 4, 8\}$ only two numbers can be selected: any of the pairs $(1, 4)$, $(1, 8)$, and $(2, 8)$;

out of $\{3, 6\}$ only one number can be selected;

out of $\{5, 10\}$ only one number can be selected;

out of $\{7\}$ only one number can be selected;

out of $\{9\}$ only one number can be selected;

Adding up all the possible choices we find that at most six numbers can form the smaller group. There are, however, as many as 12 different possibilities for selecting the numbers.

Exercise 16

Dina said the numbers $1, 15, 9, 16$, and 23 while Lila said the numbers $11, 17, 8, 5$, and 34. How many of these numbers are larger than the smallest number Lila said and smaller than the largest number Dina said?

Solution 16

The smallest number Lila said is 5. The largest number Dina said is 23. The numbers $8, 9, 11, 15, 16$, and 17 are between these limits. Therefore, there are 6 numbers that fulfill the conditions.

Exercise 17

Dina writes in increasing order a list of consecutive even numbers starting at 4. Lila writes in decreasing order a list of consecutive odd numbers starting at 19. How many numbers are there in the largest possible list of consecutive integers that can be made with numbers from both lists?

Solution 17

The largest list of consecutive integers that can be made with numbers from both lists starts at 3 and ends at 20. There are $20 - 3 + 1 = 18$ numbers on it.

Exercise 18

Amira counted how many 2-digit numbers have an even digit in the tens place and an odd digit in the units place. Lila counted how many 2-digit numbers have an odd digit in the tens place and an even digit in the units place. Who counted more numbers?

(A) Amira

(B) Lila

(C) They both counted the same number of numbers.

Solution 18

Because a 2-digit number cannot start with a zero, the tens digits that Amira can use are: 2, 4, 6, and 8. She can use any of the 5 odd digits in the units place: 1,3,5, 7, and 9.

However, Lila can use all 5 odd digits for the tens place and also 5 digits for the units place, because it is fine to have a zero in the units.

Lila will have counted more numbers than Amira.

Solutions to Miscellaneous Practice

Exercise 1

Tom alternates between eating cereal and eating muffins at breakfast. If he wants to end the month eating the same breakfast as on the first day of the month, then the month can be:

(A) April

(B) June

(C) July

(D) November

(E) any month

Solution 1

The month must have an odd number of days. April, June and November have 30 days. The only month in the list that has an odd number of days (31) is July. The answer is C.

Exercise 2

If a book has 30 pages, how many different places are there to put a bookmark in it? Assume the book has covers that are not included in the page count and begins at page one.

Solution 2

Since there are two pages for each sheet of paper (one page on each side), the book must have 15 sheets of paper, plus the two covers. There are 16 different places to put a bookmark: 14 between pages and 2 between each of the covers and the page next to it.

CHAPTER 18. SOLUTIONS TO MISCELLANEOUS PRACTICE

Exercise 3

In the *Dance of the Lilies* there are dancers dressed as pink lilies and dancers dressed as blue lilies. The blue lilies come on stage in rows of 4 dancers. How many pink lilies are there onstage, if there are 3 rows of blue lilies?

Solution 3

There are 6 pink lilies.

Exercise 4

There are five chickens in a coop. Each chicken lays an egg every other day.

1. What is the largest number of eggs the chickens can produce in one week?

2. What is the largest number of eggs the chickens can produce in two weeks?

Solution 4

1. A week has an odd number of days. The largest number of eggs a chicken can lay in a week is 4. The five chickens will lay at most 20 eggs.

2. Two weeks have an even number of days. The largest number of eggs a chicken can lay in two weeks is 7. The five chickens will lay at most 35 eggs.

Exercise 5

Max, the baker, is counting the number of people who go in and out of his shop. Starting at 10:00 am, 3 people enter and 2 people exit the shop every 2 minutes. How many more people are there in the shop at 10:10 am than at 10:00 am?

Solution 5

Every 2 minutes, the number of people in the bakery increases by one person. Within a span of 10 minutes there are 5 2-minute intervals. Therefore, there are 5 more people in the shop at 10:10 am than there were at 10:00 am.

Exercise 6

On the surface of a pond, there is a circle formed of waterlilies. On each lily there is a frog. Between any two lilies there is a frog. In total, there are 26 frogs. How many waterlilies are there?

Solution 6

In a circular pattern that alternates frogs and lilies, there are as many frogs as lilies. The number of frogs that sit on lilies is the same as the number of lilies. Therefore, the number of frogs is twice the number of lilies. There are 13 waterlilies.

Exercise 7

On a street, houses have even numbers on one side and odd numbers on the other. At one end of the street, house number 100 is across the road from house number 101. House number 202 is at the other end of the street. What is the number of the house across the road from it? (Assume all houses have the same shape and size.)

Solution 7

Make a diagram. Notice that on the odd side of the street, numbers are larger by one than the number directly opposite to them on the even side. The house number is 203:

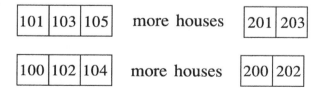

Exercise 8

Take a sheet of 8-1/2 by 11 inch letter paper and draw lines parallel to the shorter side every inch. Then, start folding along the lines in accordion style. Place the accordion on a table and notice how some creases touch the table as valleys and some point upward as hills. If the accordion has a valley at one end, does it have a valley or a hill at the other end?

Solution 8

You have drawn 9 lines and folded the paper along them. Since the number of creases (9) is odd, there must be the same type of crease at both ends. Therefore, the other end is also a valley.

Exercise 9

Cubes are used to make a square enclosure. How many cubes are needed to build an enclosure that has 8 cubes on each side?

Solution 9

We need 28 cubes. Count 4 cubes for the corners. Not counting the corners, 6 cubes are needed on each side to complete the contour:

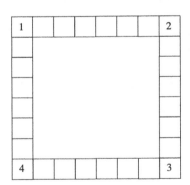

We need $6 \times 4 = 24$ cubes for the sides. Now add the corners to find the total: $24 + 4 = 28$.

Exercise 10

A cube has 6 square faces. If you place it on a table, only 5 square faces can be seen (you are allowed to walk around it). If you make a tower of 3 cubes, how many square faces are going to be visible?

Solution 10

For the top cube we can see 5 faces but for both cubes under it we can only see 4 faces since there is a cube placed on top that obstructs the top face.

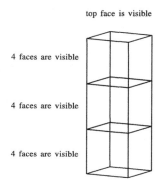

The total number of faces that can be seen is:

$$5 + 4 \times 2 = 13$$

Exercise 11

A cube has 6 square faces. If you place it on a table, only 5 square faces can be seen. If you make a tower of 10 cubes, how many square faces are going to be visible?

Solution 11

For the top cube we can see 5 faces but for each of the 9 cubes under it we can only see 4 faces since there is a cube placed on top that obstructs the top face.

In total, we can see 41 faces:

$$5 + 4 \times 9 = 5 + 36 = 41$$

Exercise 12

How many times do I have to put my right foot in front of my left foot if I take 10 steps forward?

Solution 12

5 times.

Exercise 13

Dina has 2 six sided dice with numbers from 1 to 6 on them. She throws them and adds the numbers that come up. How many possible results can she get?

Solution 13

A die has faces numbered from 1 to 6. The smallest sum is 2 and the largest sum is 12. Any number between these limits is a possible result. Therefore, there are $12 - 2 + 1 = 11$ possible results.

Exercise 14

On a six-sided die, we paint two faces black and the remaining faces red. How many different dice can we make?

Solution 14

The two black faces can be either in opposite positions or in adjacent (neighboring) positions. There are only 2 different dice possible.

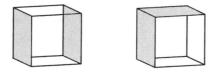

Exercise 15

Dina has 4 square pieces of paper which she can place on a grid to make a shape. The squares must share complete edges and must not overlap. The squares have the same color on both sides. How many different shapes can Dina make?

Solution 15

Five different shapes:

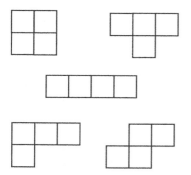

Exercise 16

Lila is making an abacus of her own design. On the first rod she threads one red bead and 11 blue beads, on the second rod she threads two red beads and 10 blue beads, and follows this pattern until the colors are completely reversed. How many rods does Lila's abacus have?

Solution 16

Strategic Solution: From row to row, one bead changes color. Since on the first row there is one red bead and 11 blue beads while on the last row there are 11 red beads and one blue bead, 10 beads must have changed color. Since there is a single change of color between rows, there must be 11 rows in total.

Brute Force Solution:

	Red beads	Blue beads
Rod 1	1	11
Rod 2	2	10
Rod 3	3	9
Rod 4	4	8
Rod 5	5	7
Rod 6	6	6
Rod 7	7	5
Rod 8	8	4
Rod 9	9	3
Rod 10	10	2
Rod 11	11	1

PRACTICE OBSERVATION

AND

LOGIC

OBSERVATION

One of the most important skills in problem solving, observation trains patience and lateral thinking.

Experiment

Which of the following figures on the right is not a larger version (possibly rotated) of the corresponding figure on the left?

Even though the pentagon and the spiral are rotated, the larger figure is still only a bigger version of the smaller figure. The small star, however, has 7 corners while the large star has 8.

If the student rushes, he or she may indicate the spiral or the pentagon as dissimilar. If the student is patient and counts the spikes of the stars, he or she will get the correct answer.

PRACTICE ONE

Exercise 1

Lila has a bracelet made of blue and white magnetic beads:

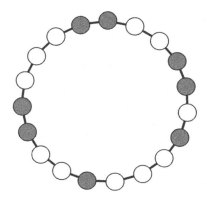

Lila would like to change the bracelet so that all beads of the same color are next to each other. She wants to do this by swapping two beads at a time. What is the smallest number of swaps she must make? (To swap means to exchange the positions of the beads in a pair.)

Exercise 2

Dina folds the following square along one of the grid lines:

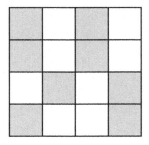

She wants to obtain the largest number of matches for the small squares after folding. A match indicates that two squares of the same color overlap each other. What is the largest number of matches she can get by cleverly choosing the line for the fold?

Exercise 3

Lila's dog lives in a rickety old doghouse. It has many holes in the roof!

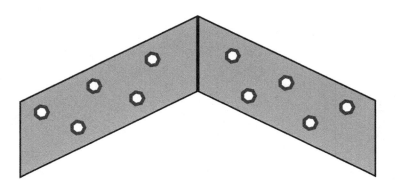

Lila wants to use some old roofing patches to fix the doghouse. Each patch looks like this:

Lila cannot cut the patches but she can, if needed, overlap them. At least how many patches must she use in order to patch all the holes?

Exercise 4

Which of the digits in the following picture is the largest?

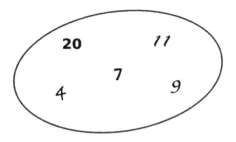

Exercise 5

Dina has a box of square tiles identical to the tile shown on the left. She uses them to build the shapes A, B, C, D, and E. Which of these shapes is made out of a different number of tiles than the other shapes?

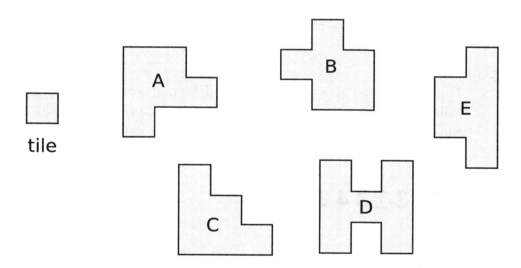

Exercise 6

How many even numbers can Lila make by removing some of the digits in the following string? Lila is not allowed to swap any of the remaining digits.

4   7   0   5

Exercise 7

There is a rule for forming the following set of numbers. In one of the numbers, some digits are missing. What could the missing digits be?

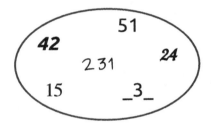

Exercise 8

Dina is allowed to swap any two adjacent (neighboring) digits in the following string. What is the smallest number of swaps she needs to make in order for all the "3"s to be one beside the other?

3 2 3 1 3 4 3

Exercise 9

Amira is turning 4 years old. Her birthday cake has been sliced into quarters with one candle for each slice, like in the diagram:

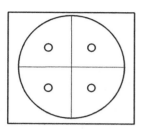

She asks her friend to move two candles at the same time and place each of them on a slice neighboring the slice it is on. What is the smallest number of such operations her friend must make in order to move all the candles to the same slice?

Exercise 10

Each of the figures in the diagram follows the same pattern. What number should be placed instead of the question mark?

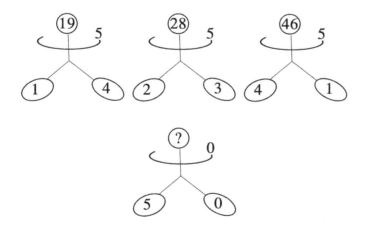

REASONING

Example:

Answer the following questions:

1. Lila has 12 candles. She cuts each candle in half. How many candles does she have now?

2. Dina has 12 matches. She cuts each match in half. How many matches does she have now?

2. Amira has 12 envelopes. She cuts each envelope in half. How many envelopes does she have now?

Answers:

Notice that the physical nature of the objects we cut in half is important.

1. Lila will have 24 smaller candles.

2. Dina will have 12 shorter matches. Since the matches have only one end that produces a flame, each cut will produce a shorter match and a simple stick.

3. Amira will have 0 envelopes left. By cutting the envelope in two, the envelope is destroyed.

PRACTICE TWO

Exercise 1

Dina is making orangeade for 10 of her friends in a large pitcher with a capacity of 8 cups. She has 3 cups of cold water in the pitcher. To this, she adds 7 cups of orange juice. How many cups of orangeade does the pitcher contain now?

Exercise 2

Lila tied her dog Arbax to a small tree using a 6 foot line. At lunch time, Lila placed some bones 10 feet away from Arbax and went to picnic with Dina nearby. Can Arbax reach the bones?

Exercise 3

Dina and Lila are throwing a party for their birthday. They have a bag of identical party balloons to inflate. Using a small pump, Lila was able to inflate a balloon in 2 minutes. Using a larger pump, Dina was able to inflate a balloon in 1 minute. Whose balloon has more air inside, Dina's or Lila's?

Exercise 4

Amira is going down the stairs. Which of her feet is on the lower step: the right one or the left one?

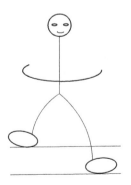

Exercise 5

Dina's phone shows the wrong time. It is 2 hours ahead of local time. If Dina's phone shows it is 11 pm, what is the actual time?

Exercise 6

Lila has three boxes and four potatoes. She places the potatoes in the boxes. Then, she makes up a number by counting the potatoes in each box. The leftmost box holds the number of hundreds, the middle box holds the number of tens, and the rightmost box holds the number of units, like in this example:

How many different numbers can she form?

Exercise 7

Lila is standing in line to buy popcorn at the cinema. There are 5 people in front of her and 3 people behind her. Each time someone buys popcorn and leaves the line, two people join the line. How many people are standing in line by the time Lila gets to order?

Exercise 8

Dina has two cups of hot chocolate milk at 90 degrees Fahrenheit. What is the temperature of each cup?

Exercise 9

When the day before yesterday was tomorrow, it was Tuesday. Tomorrow it will be:

(A) Tuesday

(B) Wednesday

(C) Thursday

(D) Friday

(E) Saturday

Exercise 10

The purple warbler has a nest full of chicks. Half of her chicks have grey heads, half of her chicks have black tails, and half of her chicks have red bellies. The number of chicks cannot be:

(A) 2

(B) 4

(C) 5

(D) 6

Exercise 11

Arbax, the Dalmatian, is 10 steps away from Dina. Dina could be:

(A) anywhere on a 20 step long segment;

(B) anywhere on a 10 step long segment;

(C) at one of the ends of a 20 step long segment;

(D) anywhere on a square with a side 10 steps long;

(E) anywhere on a circle with a radius 10 steps long.

Exercise 12

Lila and Dina have some cards marked either with '+' or with '−' and each uses the cards to make the arrangement:

$$+ - + + -$$

Then they play a game with the following the rules:

(i) replace two adjacent pluses with a minus

(ii) replace two adjacent minuses with a plus

(iii) replace two adjacent symbols that are different with a plus

until each arrangement is reduced to a single card.

Dina starts on the left of the arrangement, applies one of the rules and then starts back at the leftmost card.

Lila does the same, but she starts on the rightmost card. Which one of them has a plus left at the end of the game: Dina, Lila, or both?

Rates

A *rate* is a quantity that describes how quickly another quantity changes over *time*.

In elementary problem solving we only handle *constant* rates. We are not concerned with rates that change. This means that if a problem states that a horse travels 20 miles in one hour, we assume that the horse starts directly at a speed of 20 mph and not from rest. This is, of course, not physically correct, but it is a very good model for handling average speeds.

At this level, students must be able to differentiate a situation in which a constant rate applies from a situation in which it does not.

For problems in which constant rates do apply, simple multiplications or divisions are needed. Therefore, the students should be at least in second grade.

PRACTICE THREE

Exercise 1

Dina runs 1 mile in 10 minutes. Lila runs half a mile in 6 minutes. Which of them runs faster?

Exercise 2

Arbax, the Dalmatian, barks 3 times every 10 minutes. How many times will Arbax bark in one hour?

Exercise 3

Each minute, Dina erases 4 letters from the left while Lila erases 3 letters from the right. After how many minutes will the board be wiped clean?

kavasnexozfboiutymgafhoqetmebwletim

Exercise 4

It takes Dina 3 minutes to squeeze enough oranges to fill a glass with fresh juice. How many minutes will it take her to fill 5 glasses?

Exercise 5

It takes Lila 3 minutes to hardboil an egg. How long does it take her to hardboil 5 eggs?

Exercise 6

Two kittens can lap up a saucer of milk in 6 minutes. How long will it take four kittens to do the same?

Exercise 7

One clock is fast and advances 10 minutes more than a regular clock every hour. At 4:00 pm, it is set to show the correct time. When the faster clock is one hour ahead of it, what time will the regular clock show?

Exercise 8

One person wears one pair of pants. How many pairs of pants will two pairs of people wear?

Exercise 9

It takes Dina three days to knit a scarf and it takes Lila two days to knit a scarf. How many days do the twins need to knit 5 scarves?

Exercise 10

Dina has a magical beanstalk that triples in height each day. In six days, the beanstalk reached the height of her house: 18 feet. How many days before that was the beanstalk only 6 feet tall?

Exercise 11

Over the summer, Lila stays with her cousins while Dina stays with her grandparents. They write each other a letter every day. How many letters will they send in a week?

Exercise 12

A car is 8 times faster than a bicycle. If it takes the car 1 hour to travel a circular path, how many laps around the path can the bicycle complete in 8 hours?

LOGIC

To solve a logic problem, a student may have to make a table to summarize information.

Example There are 3 marbles, one green, one orange, and one blue, in a box. One is made of plastic and two are made of glass. The plastic one is not green. The green and the blue marbles are made of the same material. Which one is made of plastic?

	green	orange	blue
plastic	no	?	?
glass	yes	?	yes

Notice how, after completing the second line, it becomes clear that the orange marble must be made of plastic.

The student must be able to recognize *exclusive* properties and *non-exclusive* properties.

Example John, Marie, and Maggie are friends. Some of them like apples and some of them like pears. Some of them have short hair and some of them have long hair.

The sum of the number of children who have long hair and the number of children who have short hair *must* equal 3, since no one can have both long and short hair at the same time.

The sum of the number of children who like apples and the number of children who like pears, however, may exceed 3 if there are any children who like both pears and apples.

Another solving strategy is to represent several categories of objects in a diagram.

Example There are 5 finalists at a dog show. Of the finalists, 4 have long hair and 3 are good runners. At least how many are good runners with long hair?

Make a diagram separating the dogs into two groups: long-haired and short-haired.

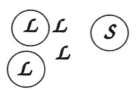

In order to find the smallest number of long-haired dogs that are good runners, we mark the short-haired dog as a good runner first. Since there are more good runners than short-haired dogs, we now mark as many long-haired dogs as possible as good runners. Don't forget, however, that there are only 3 good runners among the 5 dogs. In the following diagram, the good runners have been circled:

Thus the smallest possible number of good runners with long hair is 2.

Note: the diagram does not need to contain complicated drawings. Simple symbols will do the job.

PRACTICE FOUR

Exercise 1

The friends of a friend are my friends. This is true for all the people in my circle of friends. If each of us has 8 friends, how many of us are there?

Exercise 2

Seven frogs are sitting on some waterlilies. Suddenly, a vulture swoops down and snatches one of the frogs. How many frogs are left sitting on the waterlilies?

Exercise 3

Dina and Lila are holding the baby while their father helps the nurse set up the weighing station. At least how many people are there in the room?

Exercise 4

Dina and Lila traveled by omnibus train from Blois to Tours in 4 hours. If they had used an express train instead, the trip would only have lasted 2 hours. Which train traveled a longer distance?

Exercise 5

Dina baked a large brownie cake and cut it in 12 slices. Lila cut one of the slices in 4. How many slices are there now?

Exercise 6

Lila wants to adopt a cat. At the animal shelter there are 31 cats, of which 15 are tabby and the rest are black. The cats are so cute that Dina also decides to adopt a cat. Dina and Lila walk out, each holding the cat of her choice. What is the largest possible number of black cats that remained at the shelter after Dina and Lila left?

Exercise 7

4 race cars covered 4 miles in 1 minute. How many seconds will it take a single race car to cover 4 miles?

Exercise 8

Dina has two friends in grade 2, five friends in grade 4, and four friends in grade 3. At least how many friends does Dina have?

Exercise 9

Four of Lila's friends like Hathi, the elephant. Four of Lila's friends like Rikki, the mongoose. Four of Lila's friends like Felix, the cat. At least how many friends does Lila have?

Exercise 10

Dina has four big magnets and four small magnets. Two magnets are red, one is blue, three are green, and two are yellow. No large magnets are red. No small magnets are blue. An equal number of small and large magnets are painted in the same color. How many large green magnets are there?

Exercise 11

Lila has 28 fish in her 30 liter fish tank. One day, she removes 15 liters of water from the tank. How many fish are there in the tank now?

Exercise 12

Dina has more books than Nina. Nina has fewer books than Lila. Edna has more books than Lila. Dina does not have the largest number of books. Who has the smallest number of books?

MISCELLANEOUS PRACTICE

Exercise 1

Dina is in Ms. Left's class at the left end of the schoolyard. Lila is in Ms. Right's class at the right end of the schoolyard. In the middle of the schoolyard, there is a flag. After school, Lila runs towards Dina as Dina walks slowly towards Lila. When they meet, which one of them is closer to the flag?

Exercise 2

Dina's father is a pilot. His plane can fly at 500 miles per hour. He takes off from Plampaloosa and flies for one hour. After this time, how far away from Plampaloosa can he *not* be?

(A) 0 miles

(B) 200 miles

(C) 300 miles

(D) 500 miles

(E) 600 miles

Exercise 3

In Crazy Horse Gulch, there is an unusual weather pattern. It rains for three days, is sunny for three days, rains again for three days, and so on. If it rained today and yesterday, what will the weather be like in 4 days?

(A) rainy for sure

(B) sunny for sure

(C) could be either rainy or sunny

Exercise 4

Lila's father is 30 years older than she is. How many years older than her will her father be when Lila is 30 years old?

Exercise 5

It takes a magical fig tree 8 years to grow 10 feet tall. How many years does it take 2 magical fig trees to grow 10 feet tall?

Exercise 6

Dina and Lila harvested a lot of figs this summer and decided to dry them out in the sun. If it takes 3 days to dry 100 figs, how many days does it take to dry 300 figs?

Exercise 7

Lila and Dina ran towards each other on a 6 mile road. Dina ran twice as fast as Lila. By the time they met, how many miles had Lila run?

Exercise 8

An animation begins with a row of 25 squares, all of which are white. Every second, some of the squares turn blue. First, every second square starting from 1 turns blue. Then, every third square starting from 1 turns blue. Then, every fourth square starting from 1 turns blue, and so on. When will the entire row be blue?

Exercise 9

A *hexagon* is a closed figure with 6 sides. How many hexagons are hidden in the picture?

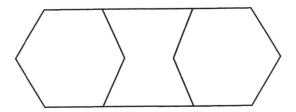

Exercise 10

Dina has lots of these tiles:

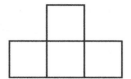

How many of the figures below can she make?

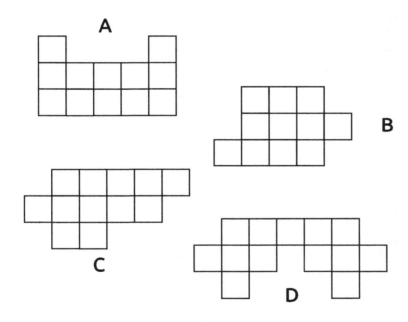

Exercise 11

Lila received a robomouse for her birthday. It was designed to walk on a grid either horizontally or vertically. Unfortunately, Lila's robomouse is defective and unable to turn left or go backwards. When Lila places it in the labyrinth at point A, facing in the direction of the arrow, it gets stuck at B.

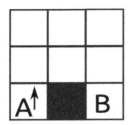

If Lila places it at point B facing in the direction of the arrow, at which of the following points will it get stuck?

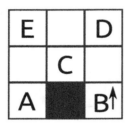

Exercise 12

Dina and Lila love archery! They shoot arrows at the following target:

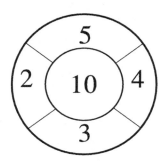

If they hit the target, they gain the number of points written on the portion of the target where the arrow landed. If a player misses the target, 3 points are deducted from her score.

Each of them takes three shots at the target and they obtain the same final score: 12 points. Both girls have scored different numbers of points at each shot. One arrow has missed the target and 5 arrows have hit the target. What is the point difference between each player's highest scoring shot?

Exercise 13

Stephan received a box with 11 pink tennis balls and 16 blue tennis balls. In his studio, he also has a box with 25 older balls, some pink and some blue. What is the largest number of pink balls he could have in total?

Exercise 14

Amira counted her dolls. Since Amira is little, she lost count a few times. She skipped 5, counted 8 twice, skipped 15 and counted 17 three times. When she was finished, she said she had 25 dolls. How many dolls does she really have?

Exercise 15

Alfonso, the grocer, has a box with 31 pineapples and mangoes. He puts another 12 pineapples and 7 mangoes in the box. There is now an equal number of pineapples and mangoes in the box. How many mangoes were there in the box before Alfonso added more fruit?

Exercise 16

Dina, Lila, and Amira had to choose their Hallowe'en costumes. They chose from "Selfless Miranda," "Shoeless Cinderella," and "Blameless Fatima." Neither Dina nor Amira were Cinderella, and neither Lila nor Dina were Miranda. Who wore "Blameless Fatima?"

Exercise 17

Amira has lots and lots of stickers with digits on them. One day, she wanted to use some of them to make all the whole numbers from 220 to 240. As she was getting ready to do so, Arbax came in and chewed up the whole stack of "7"s. Amira thought it would be a good idea to put "2"s instead of "7"s throughout. How many different numbers did she end up with?

Exercise 18

Dina places a different digit in each square and in the circle, to satisfy the comparisons. Which digit did she place in the circle?

Exercise 19

Lila played around with a drawing tool and made the rectangles in the figure:

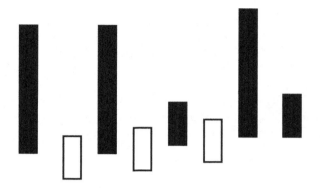

Mark the following statements as either true or false.

(A) If the bar is black, then it is tall.

(B) If the bar is tall, then it is black.

(C) If the bar is short, then it is white.

(D) If the bar is white, then it is short.

Exercise 20

Which one of the balloons illustrates the fact that "odd plus odd equals even?"

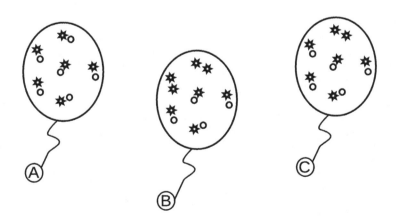

SOLUTIONS TO PRACTICE ONE

Exercise 1

Lila has a bracelet made of blue and white magnetic beads:

Lila would like to change the bracelet so that all beads of the same color are next to each other. She wants to do this by swapping two beads at a time. What is the smallest number of swaps she must make?

Solution 1

Notice how the two blue beads in the circle fit exactly in the positions occupied by the two white beads indicated by the upper arrow. Similarly, the one lonely blue bead in the triangle fits exactly in the position occupied by the white bead indicated by the lower arrow.

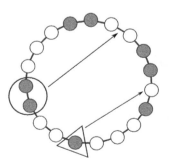

At least three swaps are needed.

Exercise 2

Dina wants to fold the square along one of the grid lines so as to obtain the largest number of matches for the small squares: black faces black and white faces white. What is the largest number of matches she can get by cleverly choosing the line for the fold?

Solution 2

If Dina folds the square along the line shown:

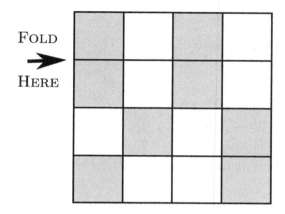

she will obtain 4 matches: 2 black and 2 white.

Exercise 3

Lila's dog lives in a rickety old doghouse. It has many holes in the roof! At least how many patches must she use in order to patch all the holes?

Solution 3

Lila notices she can cover two holes with one patch. But, because the number of holes on each slope of the roof is odd, she will have to use an extra patch. For each slope she will use 3 patches for a total of 6 patches.

Exercise 4

Which of the digits in the following picture is the largest?

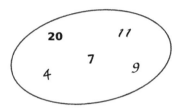

Solution 4

The question is about *digits* not about *numbers*! The largest digit used is 9.

Exercise 5

Dina has a box of square tiles identical to the tile shown on the left. She uses them to build the shapes A, B, C, D, and E. Which of these shapes is made out of a different number of tiles than the other shapes?

Solution 5

Subdivide each shape into tiles and count the number of tiles for each. A, B, C, and E are constructed out of 6 tiles while D is constructed out of 7 tiles.

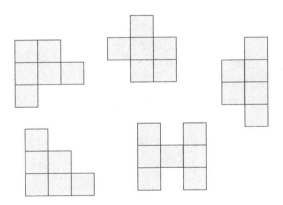

Exercise 6

How many even numbers can Lila make by removing some of the digits in the following string? Lila is not allowed to swap any of the remaining digits.

4   7   0   5

Solution 6

Lila can form the even numbers:

4, 0, 40, 70, and 470

for a total of 5 even numbers.

Exercise 7

There is a rule for forming the following set of numbers. In one of the numbers, some digits are missing. What could the missing digits be?

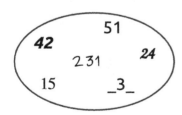

Solution 7

Notice that there are two possible solutions:

(i) The numbers form pairs which have the same digits but in reversed order. The missing number could be 132.

(ii) All the numbers have a sum of digits of 6. The missing number could be one of: 330, 132, and 231.

Exercise 8

Dina is allowed to swap any two adjacent (neighboring) digits in the following string. What is the smallest number of swaps she needs to make in order for all the "3"s to be one beside the other?

Solution 8

Dina can solve this problem in several ways. One possible way is:

3 2 3 1 3 4 3
2 3 3 1 3 4 3
2 3 3 **3** 1 4 3
2 3 3 3 1 **3 4**
2 3 3 3 **3** 1 4

which consists of 4 swaps.

Exercise 9

Amira asks her friend to move two candles at the same time and place each of them on a slice neighboring the slice it is on. What is the smallest number of such operations her friend must make in order to move all candles to the same slice?

Solution 9

Two moves are sufficient to place all the candles on the same slice:

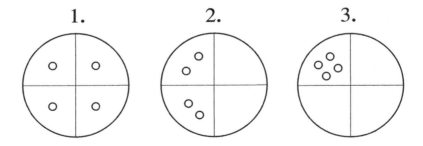

Exercise 10

Each of the figures in the diagram follows the same pattern. What number should be placed instead of the question mark?

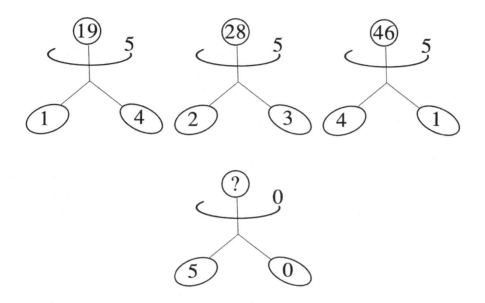

Solution 10

The pattern is to make a two-digit number using the digits placed on the "feet" and add to this number the digit placed in the "hand." The answer is 50.

SOLUTIONS TO PRACTICE TWO

Exercise 1

Dina is making orangeade for 10 of her friends in a large pitcher with a capacity of 8 cups. She has 3 cups of cold water in the pitcher. To this, she adds 7 cups of orange juice. How many cups of orangeade does the pitcher contain now?

Solution 1

Since the pitcher has a capacity of 8 cups, it cannot contain more orangeade than this maximum amount. Dina either spilled some juice or stopped pouring when she noticed the pitcher was full. The number of friends is not relevant.

Exercise 2

Lila has tied her dog Arbax to a small tree using a 6 foot line. When it was time for lunch, Lila placed some bones 10 feet away from Arbax, and went to picnic with Dina nearby. Can Arbax reach the bones?

Solution 2

The bones can be more than 6 feet from Arbax. As long as the bones are within 6 feet of the pole, Arbax will be able to reach them. This is possible in many ways, including this one:

Exercise 3

Dina and Lila are giving a party for their birthday. They have a bag of party balloons to inflate. Using a small pump, Lila was able to inflate a balloon in 2 minutes. Using a larger pump, Dina was able to inflate a balloon in 1 minute. Which balloon has more air inside?

Solution 3

The two balloons are identical and have approximately the same amount of air inside, regardless of how they were inflated.

Exercise 4

Amira is going down the stairs. Which of her feet is on the lower step: the right one or the left one?

Solution 4

Amira's left foot is on the lower step.

Exercise 5

Dina's phone shows the wrong time. It is 2 hours ahead of local time. If Dina's phone shows it is 11 pm, what is the actual time?

Solution 5

The local time is 9 pm.

Exercise 6

Lila has three boxes and four potatoes. She places the potatoes in the boxes. Then, she makes up a number by counting the potatoes in each box. The leftmost box holds the number of hundreds, the middle box holds the number of tens, and the rightmost box holds the number of units. How many different numbers can she form?

Solution 6

Lila can form the following numbers:

All potatoes are in the rightmost box: 4

All potatoes are in the middle box: 40.

All potatoes are in the leftmost box: 400.

Only the leftmost box is empty: 13, 22, 31.

Only the middle box is empty: 103, 301, 202.

Only the rightmost box is empty: 130, 220, 310.

There is at least one potato in each box: 112, 121, 211.

Note to parents: This problem helps students gain hands-on experience with counting the number of ways n objects can be put into $m < n$ boxes.

Exercise 7

Lila is standing in line to buy popcorn at the cinema. There are 5 people in front of her and 3 people behind her. Each time someone buys popcorn and leaves the line, two people join the line. How many people are standing in line by the time Lila gets to order?

Solution 7

14 people: Lila, the 3 people who were behind her at the start of the problem, and the 10 people who lined up behind her while she waited.

Exercise 8

Dina has two cups of hot chocolate milk at 90 degrees Fahrenheit. What is the temperature of each cup?

Solution 8

Each cup has a temperature of 90 degrees Fahrenheit. Temperatures are not added together.

Exercise 9

When the day before yesterday was tomorrow, it was Tuesday. Tomorrow it will be:

(**A**) Tuesday

(**B**) Wednesday

(**C**) Thursday

(**D**) Friday

(**E**) Saturday

Solution 9

Place *today* on a timeline and model the information in the problem on the timeline.

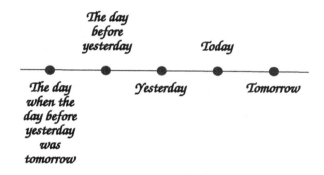

Then, simply mark the days of the week on the timeline in order:

to find out that it will be Saturday.

Exercise 10

The purple warbler has a nest full of chicks. Half of her chicks have grey heads, half of her chicks have black tails, and half of her chicks have red bellies. The number of chicks cannot be:

(A) 2

(B) 4

(C) 5

(D) 6

Solution 10

Since the number of chicks must have an integer half, she cannot have a total of 5 chicks. All other choices are possible answers.

Exercise 11

Arbax, the Dalmatian, is 10 steps away from Dina. Dina could be:

(A) anywhere on a 20 step long segment;

(B) anywhere on a 10 step long segment;

(C) at one of the ends of a 20 step long segment;

(D) anywhere on a square with a side 10 steps long;

(E) anywhere on a circle with a radius 10 steps long.

Solution 11

Dina could be anywhere on a circle with a radius 10 steps long.

Exercise 12

Lila and Dina have some cards marked either with '+' or with '−' and each uses the cards to make the arrangement:

$$+ - + + -$$

Then they play a game with the following the rules:

(i) replace two adjacent pluses with a minus

(ii) replace two adjacent minuses with a plus

(iii) replace two adjacent symbols that are different with a plus

until each arrangement is reduced to a single card.

Dina starts on the left of the arrangement, applies one of the rules and then starts back at the leftmost card. Lila does the same, but she starts on the rightmost card. Which one of them has a plus left at the end of the game: Dina, Lila, or both?

Solution 12

Dina ends up with a plus and Lila ends up with a minus. Dina's moves:

$+ + + -$

$- + -$

$+ -$

$+$

Lila's moves:

$+ - + +$

$+ - -$

$+ +$

$-$

SOLUTIONS TO PRACTICE THREE

Exercise 1

Dina runs 1 mile in 10 minutes. Lila runs half a mile in 6 minutes. Which of them runs faster?

Solution 1

If Dina runs 1 mile in 10 minutes, then she runs half a mile in 5 minutes. Lila runs half a mile in 6 minutes. By comparison, Dina runs faster.

Exercise 2

Arbax, the Dalmatian, barks 3 times every 10 minutes. How many times will Arbax bark in one hour?

Solution 2

Arbax barks 3 times every hour and 3 times every 10, 20, 30, 40, and 50 minutes after the hour. He barks 18 times during one hour.

Exercise 3

Each minute, Dina erases 4 letters from the left while Lila erases 3 letters from the right. After how many minutes will the board be wiped clean?

kavasnexozfboiutymgafhoqetmebwletim

Solution 3

Count all the letters - there are 35 of them. Each minute, Dina and Lila together erase a total of 7 letters. It takes them 5 minutes to wipe the board clean.

Exercise 4

It takes Dina 3 minutes to squeeze enough oranges to fill a glass with fresh juice. How many minutes will it take her to fill 5 glasses?

Solution 4

15 minutes

Exercise 5

It takes Lila 3 minutes to hardboil an egg. How long does it take her to hardboil 5 eggs?

Solution 5

3 minutes. Lila can boil all the eggs in the same pot.

Exercise 6

Two kittens can lap up a saucer of milk in 6 minutes. How long will it take four kittens to do the same?

Solution 6

Four kittens will drink the milk faster. They will empty the saucer in 3 minutes.

Exercise 7

One clock is fast and advances 10 minutes more than a regular clock every hour. At 4:00 pm, it is set to show the correct time. When the faster clock is one hour ahead of it, what time will the regular clock show?

Solution 7

It will take 6 hours for the faster clock to be one hour ahead of the regular clock. In six hours' time, it will be 10 pm.

Exercise 8

One person wears one pair of pants. How many pairs of pants will two pairs of people wear?

Solution 8

A pair of people is 2 people. Two pairs of people are 4 people. Each person wears one pair of pants. The answer is: 4 pairs of pants.

Exercise 9

It takes Dina three days to knit a scarf and it takes Lila two days to knit a scarf. How many days do the twins need to knit 5 scarves?

Solution 9

Dina knits two scarves in six days. Lila knits three scarves in six days. It takes six days for the girls to knit five scarves.

Exercise 10

Dina has a magical beanstalk that triples in height each day. In six days, the beanstalk reached the height of her house: 18 feet. How many days before that was the beanstalk only 6 feet tall?

Solution 10

Go backwards in time from day to day, dividing the height by 3 each time. Since $18 \div 3 = 6$, the beanstalk must have been 6 feet tall only the day before.

Exercise 11

Over the summer, Lila stays with her cousins while Dina stays with her grandparents. They write each other a letter every day. How many letters will they send in a week?

Solution 11

Dina sends Lila 7 letters, one for each day of the week. Lila also sends 7 letters during that week. In total, they send 14 letters in a week.

Exercise 12

A car is 8 times faster than a bicycle. If it takes the car 1 hour to travel a circular path, how many laps can the bicycle complete in 8 hours?

Solution 12

One lap.

Solutions to Practice Four

Exercise 1

The friends of a friend are my friends. This is true for all the people in my circle of friends. If each of us has 8 friends, how many are we?

Solution 1

Since I have 8 friends, there are at least 9 of us. Since this is true for each of us, there are exactly 9 of us.

Exercise 2

Seven frogs are sitting on some waterlilies. Suddenly, a vulture swoops down and snatches one of the frogs. How many frogs are left sitting on the waterlilies?

Solution 2

None. The ones spared by the vulture got scared and scampered off.

Exercise 3

Dina and Lila are holding the baby while their father helps the nurse set up the weighing station. At least how many people are there in the room?

Solution 3

At least five people: Dina, Lila, their father, the nurse, and the baby.

Exercise 4

Dina and Lila traveled by omnibus train from Blois to Tours in 4 hours. If they had used an express train instead, the trip would only have lasted 2 hours. Which train traveled a longer distance?

Solution 4

Both vehicles traveled the same distance: from Blois to Tours.

Exercise 5

Dina baked a large brownie cake and cut it in 12 slices. Lila cut one of the slices in 4. How many slices are there now?

Solution 5

15 slices. Three more slices were added when Lila cut one of the slices in four.

Exercise 6

Lila wants to adopt a cat. At the animal shelter there are 31 cats, of which 15 are tabby and the rest are black. The cats are so cute that Dina also decides to adopt a cat. Dina and Lila walk out, each holding the cat of her choice. What is the largest possible number of black cats that remained at the shelter after Dina and Lila left?

Solution 6

At the shelter there were 16 black cats to start with:

$$31 - 15 = 16$$

Since both Dina and Lila may have chosen tabby cats, the largest possible number of black cats remaining is 16.

Exercise 7

4 race cars covered 4 miles in 1 minute. How many seconds will it take a single race car to cover 4 miles?

Solution 7

Still 1 minute.

Exercise 8

Dina has two friends in grade 2, five friends in grade 4, and four friends in grade 3. At least how many friends does Dina have?

Solution 8

Dina has at least $2 + 5 + 4 = 11$ friends, since none of her friends can be in two different grades simultaneously.

Exercise 9

Four of Lila's friends like Hathi, the elephant. Four of Lila's friends like Rikki, the mongoose. Four of Lila's friends like Felix, the cat. At least how many friends does Lila have?

Solution 9

Lila has at least 4 friends. Each friend might like all the animals.

Exercise 10

Dina has four big magnets and four small magnets. Two magnets are red, one is blue, three are green, and two are yellow. No large magnets are red. No small magnets are blue. An equal number of small and large magnets are painted in the same color. How many large green magnets are there?

Solution 10

The only colors that even numbers of magnets are painted in are red and yellow. Therefore, the group containing the equal number of small and large magnets must have one of these two colors. Since no large magnets are red, yellow is the only option. Hence, one large magnet is yellow and one small magnet is yellow. The blue magnet must be large. There are two large magnets left unaccounted for and they can only be green. There are two large green magnets (and one small green magnet).

Exercise 11

Lila has 28 fish in her 30 liter fish tank. One day, she removes 15 liters of water from the tank. How many fish are there in the tank now?

Solution 11

There will still be 28 fish.

Exercise 12

Dina has more books than Nina. Nina has fewer books than Lila. Edna has more books than Lila. Dina does not have the largest number of books. Who has the smallest number of books?

Solution 12

Make a diagram showing the number of books each girl has in decreasing order. The data is not sufficient to determine this order completely, but may be sufficient to find the answer. Mark the positions that are not completely determined with a question mark:

$$E \qquad D \qquad L \qquad N$$
$$ \qquad ? \qquad ? \qquad $$

Dina and Lila could be interchanged. Since Edna has more books than Lila and Dina does not have the largest number of books, it follows that Edna must have the largest number of books. Nina, however, has fewer books than any of the others.

SOLUTIONS TO MISCELLANEOUS PRACTICE

Exercise 1

Dina is in Ms. Left's class at the left end of the schoolyard. Lila is in Ms. Right's class at the right end of the schoolyard. In the middle of the schoolyard, there is a flag. After school, Lila runs towards Dina as Dina walks slowly towards Lila. When they meet, which one of them is closer to the flag?

Solution 1

When they meet, they are the same distance from the flag.

Note to parents: Students must be familiarized with common approximations that are used in word problems. One of them is illustrated by this example: moving objects are generally assumed to be *pointlike*, unless otherwise specified. That is, they do not have a size of their own. Otherwise, we could rightly say that Lila is slightly closer.

Exercise 2

Dina's father is a pilot. His plane can fly at 500 miles per hour. He takes off from Plampaloosa and flies for one hour. After this time, how far away from Plampaloosa can he *not* be?

(A) 0 miles

(B) 200 miles

(C) 300 miles

(D) 500 miles

(E) 600 miles

Solution 2

The plane could not have travelled 600 miles. Notice that the plane can fly upwards, as well as turn back. All distances smaller than or equal to the 500 mile range are possible.

Exercise 3

In Crazy Horse Gulch, there is an unusual weather pattern. It rains for three days, is sunny for three days, rains again for three days, and so on. If it rained yesterday and today, what will the weather be like in 4 days?

(A) rainy for sure

(B) sunny for sure

(C) could be either rainy or sunny

Solution 3

As the diagram shows, there are two possible cases:

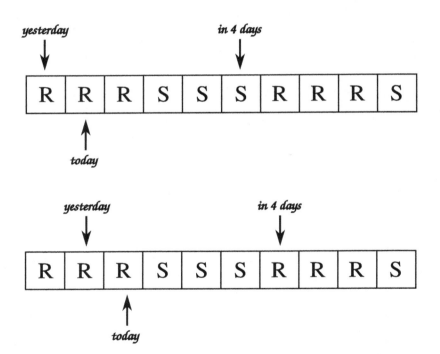

Therefore, it could be either rainy or sunny in four days.

Exercise 4

Lila's father is 30 years older than she is. How many years older than her will her father be when Lila is 30 years old?

Solution 4

The difference in the ages of two people is the same throughout their lives. Lila's father will still be 30 years older than her.

Exercise 5

It takes a magical fig tree 8 years to grow 10 feet tall. How many years does it take 2 magical fig trees to grow 10 feet tall?

Solution 5

Still 8 years. Each tree grows on its own.

Exercise 6

Dina and Lila harvested a lot of figs this summer and decided to dry them out in the sun. If it takes 3 days to dry 100 figs, how many days does it take to dry 300 figs?

Solution 6

3 days. It takes 300 figs just as long to dry as it takes 100 figs.

Exercise 7

Lila and Dina ran towards each other on a 6 mile road. Dina ran twice as fast as Lila. By the time they met, how many miles had Lila run?

Solution 7

Dina ran 4 miles and Lila ran 2 miles until they met $(2 + 4 = 6)$.

Exercise 8

An animation begins with a row of 25 squares, all of which are white. Every second, some of the squares turn blue. First, every second square starting from 1 turns blue. Then, every third square starting from 1 turns blue. Then, every fourth square starting from 1 turns blue, and so on. When will the entire row be blue?

Solution 8

None of the rules makes the second square change color. It will remain white indefinitely.

Exercise 9

A *hexagon* is a closed figure with 6 sides. How many hexagons are hidden in the picture?

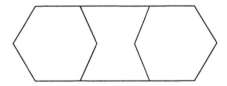

Solution 9

We count 6 hexagons:

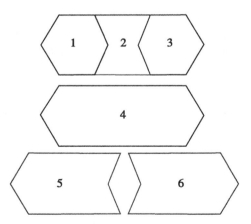

Exercise 10

Dina has lots of these tiles:

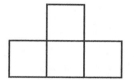

How many of the figures below can she make?

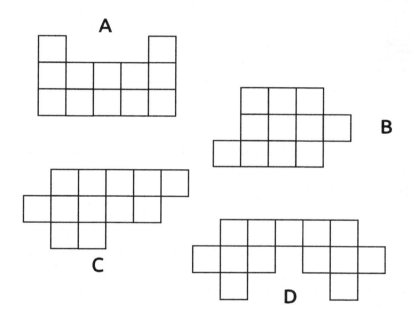

Solution 10

Figures B and D cannot be constructed using Dina's tiles. Only A and C can be built.

Exercise 11

Lila received a robomouse for her birthday. It was designed to walk on a grid either horizontally or vertically. Unfortunately, Lila's robomouse is defective and unable to turn left or go backwards. When Lila places it in the labyrinth at point A, facing in the direction of the arrow, it gets stuck at B. If Lila places it at point B facing in the direction of the arrow, at which of the following points will it get stuck?

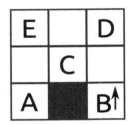

Solution 11

The robomouse will get stuck at D since the only moves possible at D would be to turn left or to go back. The robomouse is defective and cannot turn left or go back.

Exercise 12

Dina and Lila love archery! They shoot arrows at the following target:

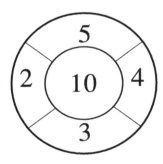

If they hit the target, they gain the number of points written on the portion of the target where the arrow landed. If a player misses the target, 3 points are deducted from her score.

Each of them takes three shots at the target and they obtain the same final score: 12 points. Both girls have scored different numbers of points at each shot. One arrow has missed the target and 5 arrows have hit the target. What is the point difference between each player's highest scoring shot?

Solution 12

It is possible to score 12 points in three shots in one of the three following ways:

$$4 + 4 + 4 = 12$$
$$4 + 5 + 3 = 12$$
$$10 + 5 - 3 = 12$$
$$5 + 5 + 2 = 12$$

However, two of the possibilities have repeated scores. Or, the problem says that each of the girls scored a different number of points on each shot. Therefore, the only valid possibilies are:

$$4 + 5 + 3 = 12$$
$$10 + 5 - 3 = 12$$

The highest scores are 10 and 5. The difference between them is 5.

Exercise 13

Stephan received a box with 11 pink tennis balls and 16 blue tennis balls. In his studio, he also has a box with 25 older balls, some pink and some blue. What is the largest number of pink balls he could have in total?

Solution 13

Since we know for sure that in the old box not all the balls are pink, there are at most 24 pink balls in it. Add the 11 new pink balls to find the largest possible number of pink balls: $24 + 11 = 35$.

Exercise 14

Amira counted her dolls. Since Amira is little, she lost count a few times. She skipped 5, counted 8 twice, skipped 15 and counted 17 three times. When she was finished, she said she had 25 dolls. How many dolls does she really have?

Solution 14

Skipping a number is like counting a doll she does not have. For instance, if she counts "second doll, fourth doll," she actually has only 3 dolls at this point. Since she skips two numbers, we must decrease her total by 2: $25 - 2 = 23$.

Counting a number twice means that Amira has a doll but fails to account for it and gives it the same number as the previous doll. Therefore, we must add 1 to the total: $23 + 1 = 24$.

Counting a number three times means that Amira has two dolls she fails to account for and gives the same number as the previous doll. Therefore, we must add 2 to the total: $24 + 2 = 26$.

Amira has 26 dolls.

Exercise 15

Alfonso, the grocer, has a box with 31 pineapples and mangoes. He puts another 12 pineapples and 7 mangoes in the box. There is now an equal number of pineapples and mangoes in the box. How many mangoes were there in the box before Alfonso added more fruit?

Solution 15

Add all the fruit together to find the total: $31 + 12 + 7 = 50$. Of these, half (25) are pineapples and half are mangoes. If Alfonso had to add 7 mangoes to reach the count of 25, then there were $25 - 7 = 18$ mangoes in the box before he added more fruit.

Exercise 16

Dina, Lila, and Amira had to choose their Hallowe'en costumes. They chose from "Selfless Miranda," "Shoeless Cinderella," and "Heartless Fatima." Neither Dina nor Amira were Cinderella, and neither Lila nor Dina were Miranda. Who wore "Blameless Fatima?"

Solution 16

If Dina and Amira are not Cinderella, then Lila must be Cinderella. If Lila and Dina are not Miranda, then Amira must be Miranda. This leaves Dina as Fatima.

Exercise 17

Amira has lots and lots of stickers with digits on them. One day, she wanted to use some of them to make all the whole numbers from 220 to 240. As she was getting ready to do so, Arbax came in and chewed up the whole stack of "7"s. Amira thought it would be a good idea to put "2"s instead of "7"s throughout. How many different numbers did she end up with?

Solution 17

Without Arbax's intervention, Amira would have created $240 - 220 + 1 = 21$ numbers. Because of the change, 227 is turned into 222 and 237 is turned into 232. 222 and 232 are numbers in the set that Amira wants to create. Therefore, they will appear twice. The number of different numbers is $21 - 2 = 19$.

Exercise 18

Dina places a different digit in each square and in the circle, to satisfy the comparisons. Which digit did she place in the circle?

Solution 18

The total number of squares and circles is 10. This means that all 10 digits must be used. Though there are many ways of placing digits in the squares, only zero can be placed in the circle. We show two different solutions in the figures. From these, one can see how more solutions can be built.

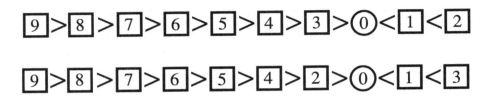

Regardless of how we arrange the digits in the squares, the digit in the circle has to be smaller than all other digits.

Exercise 19

Lila played around with a drawing tool and made the rectangles in the figure:

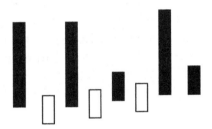

Mark the following statements as either true or false.

(A) If the bar is black, then it is tall.
(B) If the bar is tall, then it is black.
(C) If the bar is short, then it is white.
(D) If the bar is white, then it is short.

Solution 19

(A) False. Some black bars are short.
(B) True. All tall bars are black.
(C) False. Some short bars are black.
(D) True. All white bars are short.

Exercise 20

Which one of the balloons illustrates the fact that "odd plus odd equals even?"

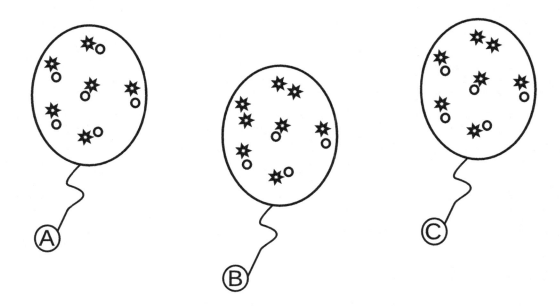

Solution 20

Since it is possible to pair the objects in each balloon, the total number of objects in each balloon is even. The three balloons illustrate the two different cases of addition that can produce an even sum:

(A) $6 + 6 = 12$, "Even plus even equals even."

(B) $8 + 4 = 12$, "Even plus even equals even."

(C) $5 + 7 = 12$, "Odd plus odd equals even."

The answer is (C).

PRACTICE ARITHMETIC

DIGITS AND NUMBERS

Digits are symbols we use to write out numbers. In the decimal system there are 10 digits available: 0, 1, 2, 3, 4, 5, 6, 7, 8, and 9.

Numbers are written using digits. For example, 23 is a 2-digit number, 6 is a one-digit number, and 50031 is a 5-digit number.

0, 2, 4, 6, and 8 are *even* digits.

1, 3, 5, 7, and 9 are *odd* digits.

Non-zero digits range from 1 to 9.

The *number of digits* of a number is often used in problems. If the number of digits is given, then the leftmost digit must be a non-zero digit.

The *first* digit of a number is always the leftmost digit of the number. The first digit of a number must be a non-zero digit.

The *last* digit of a number is always the rightmost digit of the number.

The *digit sum* of a number is simply what we get if we add all the digits.

Example The digit sum of the number 421 is $4 + 2 + 1 = 8$.

The *digit product* of a number is simply what we get if we multiply all the digits.

Example The digit product of the number 421 is $4 \times 2 \times 1 = 8$.

Because digits range from 0 to 9, interesting observations can be made about the size of the carryover when we add or multiply numbers.

For example, the *only* carryover possible when adding two digits is 1, because two digits add up to at most 18. Even if two 9s are added and there is a carryover of 1 from the previous place value, the carryover will still not exceed 1.

One can find the largest possible carryover when adding three or more digits using a similar reasoning.

Experiment

1. Write the smallest odd numbers with 4, 5, and 6 digits.

2. Write the smallest 3-digit number with odd digits that are all different.

3. Write the largest 4-digit number with a digit sum equal to 3.

4. Write a 6 digit number so that the product of its digits is equal to 1.

5. Write the smallest 4-digit number that can be written using the digits 0, 3, and 7 at least one time each.

6. Write the largest number that contains each odd digit once.

7. Write the smallest number that contains all the even digits.

Answers:

1. 1001, 10001, and 100001

2. 135

3. 3000

4. 111111

5. 3007

6. 97531

7. 20468

PRACTICE ONE

Exercise 1

Make a list of all the numbers from 1 to 30 that have only even digits.

Exercise 2

What is the largest even 3-digit number?

Exercise 3

What is the largest 3-digit number with all digits even?

Exercise 4

The figure below is painted on the side of a truck. If the truck has stopped beside a lake, how many digits can be seen in its reflection in the water?

0123456789

Exercise 5

We have the following 2 cards:

86   **69**

How many different numbers can we make using them? The cards are square and may be rotated, but they are opaque (not see-through.)

Exercise 6

If we add two digits, what is the largest carryover we can get?

(A) 1

(B) 2

(C) 9

(D) any number

Exercise 7

If we add three digits, what is the largest carryover we can get?

Exercise 8

Lila adds two digits and gets a carryover of 1. One of the digits is smaller than 5. The other digit is:

(A) larger than or equal to 5

(B) larger than 5

(C) smaller than 5

(D) any digit

Exercise 9

Dina tries to write the number 121212121212 as the sum of two numbers that only have digits of 1 and 0. Can you help her?

Exercise 10

Which is the larger sum?

```
    111111        123456
    222222        234561
    333333        345612
    444444        456123
    555555        561234
    666666        612345
    ────────  +   ────────  +
```

Exercise 11

> If you count backwards from 999 to 1, at which count will the digit in the middle change for the first time?

Exercise 12

> How many three digit numbers can be formed using only the digits 7 and 0?

Exercise 13

> Make a list of the digit sums for all the numbers between 11 and 20.

Exercise 14

> Lila made a list of all the numbers smaller than 100 that have a digit sum of 16. How many different digits did she use while writing the list?

Exercise 15

> How many 5-digit numbers have a digit sum of 2?

Exercise 16

> Dina has written down a 6 digit number. Lila reverses the digits of this number. Then, the two girls compare their numbers and find that they are identical. What is the largest number of different digits Dina's number could have?

Exercise 17

> Lila asks Dina: "For how many even 3-digit numbers does the digit sum equal 26?" Dina finds a number. Can you help her find more?

Exercise 18

> The number 333 is written using only the digit 3. In how many different ways can we write it as a sum of two positive numbers that are written using only the digits 3 and zero?

Exercise 19

Dina made a list of all the numbers written by repeating the same digit that are greater than 405678 and smaller than 999110. Lila counted the numbers on Dina's list. How many numbers did Lila count?

Exercise 20

Dina thinks of a number and Lila thinks of a number. When they add their numbers, the sum is even. If they subtract the smaller number from the larger number, is the difference even or odd?

Exercise 21

If Lila gave Dina 5 beads, they would both have the same number of beads. What is the difference between the number of beads Lila has and the number of beads Dina has?

Exercise 22

If Lila gave Dina 5 beads, they would both have the same number of beads. Which of the following could not be the total number of beads they have?

(A) 21 beads

(B) 22 beads

(C) 24 beads

(D) 104 beads

Exercise 23

What is the largest possible digit sum for a 3-digit number in which all the digits are different?

Exercise 24

A computer begins counting by ones starting from 13579. What is the next number with odd digits that are all different?

Exercise 25

Arbax, the Dalmatian, has 16 bones hidden in 5 different caches. Arbax thinks there is an odd number of bones in each cache. Is Arbax right?

CRYPTARITHMS

A *cryptarithm* is a mathematical riddle. It consists of a simple operation, such as an addition or a multiplication, in which some or all the digits are replaced by symbols (*encrypted*).

The rules of the cryptarithm are often part of the problem statement and are generally, but not always, as follows:

- different symbols represent different digits,
- the same symbol represents the same digit.

Cryptarithms can be solved using a variety of techniques. Among these techniques, the following are the most important:

- no number ever starts with a 0,
- digits range from 0 to 9 only,
- the carryover is limited to certain values, based on the operation and the number of digits,
- digits already discovered can be ruled out for the remaining symbols,
- the parity of the digits (even or odd) is sometimes predictable,
- if it is possible to bracket the values of a symbol within narrow limits (for example, we conclude that some symbol can only be 1 or 3) it is usually a good idea to try out each value in turn.

Experiment

Try this very simple cryptarithm:

$$\heartsuit\spadesuit + \spadesuit = \clubsuit\diamondsuit\diamondsuit$$

Because the result is a three digit number, its leftmost digit can only be a 1 that has been carried over from the previous place value.

This carryover can only be obtained if the digit marked as a $\heartsuit$ is a 9 and there is a carryover of 1 from the previous place value.

The $\clubsuit\diamondsuit$, therefore, represents the number 10.

Adding the two $\spadesuit$ symbols must produce a carryover and a last digit of zero. Therefore, the $\spadesuit$ must be a 5, and the addition is:

$$95 + 5 = 100$$

When solving cryptarithms, it is important to work logically and not by trial-and-error. The student should work from clue to clue until the problem is completely decrypted.

PRACTICE TWO

Exercise 1

In the following figure, different shapes represent different digits, and the same shape represents the same digit. Which digit does the square represent?

Exercise 2

In the following sum, different letters represent different digits:

$$A + A + A + A + A = B$$

Find the digits!

Exercise 3

If A and B are digits and $A + A = B$, how many different values can A have?

Exercise 4

Two different digits cannot have a difference of:

(A) 0

(B) 3

(C) 5

(D) 9

Exercise 5

In the following figure, different shapes represent different digits and the same shape represents the same digit. Which digit does the square represent?

Exercise 6

Two numbers are encoded as *AC* and *BC*, where different letters represent different digits and the same letter always represents the same digit. Their sum could be any of the following, except:

(A) 100

(B) 102

(C) 103

(D) 104

Exercise 7

Each symbol represents a digit from 0 to 9. Different symbols represent different digits. Find out what the digits are. How many such additions are there?

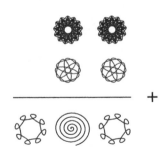

Exercise 8 Lila has found a solution to the following cryptarithm. Arbax, the Dalmatian, is also working hard to find a solution. Can he find a solution different from Lila's?

$$
\begin{array}{r}
\mathbf{XOX} \\
\mathbf{XOX} \\
\hline
\mathbf{OHH}
\end{array} +
$$

Exercise 9 Which digits do the symbols represent?

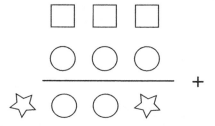

Exercise 10 Each letter represents a different digit. The same letter is always the same digit. Find the digit that corresponds to each letter:

$$
\begin{array}{r}
\mathbf{D} \\
\mathbf{D}\ \mathbf{D} \\
\mathbf{D}\ \mathbf{D}\ \mathbf{D} \\
\hline
\mathbf{A}\ \mathbf{A}\ \mathbf{B}\ \mathbf{C}
\end{array} +
$$

Exercise 11 In the following examples, O represents an odd digit and E represents an even digit. Which example is impossible?

$$
\begin{array}{cc}
O & E \\
E & O \\
\hline
E & O
\end{array}
\ + \ \text{(A)}
\qquad\qquad
\begin{array}{cc}
E & E \\
O & O \\
\hline
O & O
\end{array}
\ + \ \text{(C)}
$$

$$
\begin{array}{cc}
O & E \\
E & E \\
\hline
O & O
\end{array}
\ + \ \text{(B)}
\qquad\qquad
\begin{array}{cc}
O & E \\
O & O \\
\hline
E & O
\end{array}
\ + \ \text{(D)}
$$

Exercise 12

In the following correct additions, the same symbol represents the same digit and different symbols represent different digits. Which sum is different from all others?

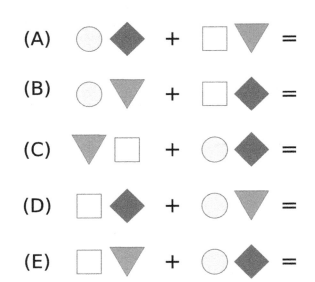

EQUATIONS

Solving problems using *equations* can be introduced at this age level by using concrete representations for the quantities involved. To solve problems using a concrete technique, use the following steps:

- read the problem carefully,
- identify the quantity that is the most basic and model it using a rectangle,
- use the same rectangle for the same amount,
- observe how the rectangles must be combined,
- figure out which arithmetic operation models the required combination.

Example

Dina had 4 more dinosaurs than Lila. Lila received 16 more dinosaurs for her birthday and now she has twice as many dinosaurs as Dina. How many dinosaurs did Lila have before her birthday?

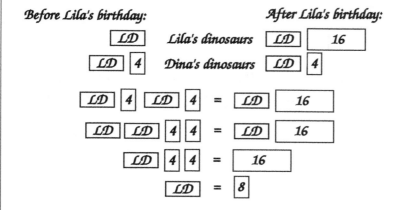

Example

A baker has made twice as many baguettes as loaves and half as many rolls as loaves. If he now has 140 products up for sale, how many baguettes has he made?

Find the product with the smallest amount. It's the rolls, right? Make a rectangle that represents the number of rolls:

Rolls

Read the problem again. There are twice as many loaves as rolls.

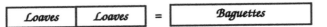

Read the problem again. There are twice as many baguettes as loaves.

Loaves	*Loaves*	=	*Baguettes*

Now make a diagram with all the products using the number of rolls as a unit:

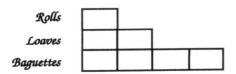

There are 7 rectangles. If there are 140 products in total, each rectangle represents 20 products. Therefore, he baked 20 rolls, 40 loaves, and 80 baguettes.

168

PRACTICE THREE

Exercise 1

Dina walked 6 steps up from the bottom of a staircase while Lila walked 5 steps down from the top. They met on the same step. How many steps does the staircase have?

Exercise 2

When Arbax, the Dalmatian, celebrates his 7th birthday, Lila will be 16. Today, it is Arbax's birthday and Lila is 10. How old is Arbax?

Exercise 3

Dina's mother is cooking stuffed peppers. For each pepper, she chops one tomato and two scallions. She fills a tray with 6 stuffed peppers. How many vegetables has she used in total?

Exercise 4

There are four chairs at the breakfast counter in Lila's kitchen. At breakfast, Dina, Lila, and their parents sit down for their meal. How many legs can one count?

Exercise 5

On a good spring day, one can see deer and hares from Dina's porch. Today, Dina saw as many hare's ears as deer's tails. If Dina saw 7 hares, how many deer did she see?

Exercise 6

Lila gave Amira 12 magic wands. Amira gave Dina 8 magnets. Dina gave Lila 14 crystals. Which girl experienced the largest difference in her total number of toys? Was it an increase or a decrease?

Exercise 7

Arbax, the Dalmatian, had stashed away 25 bones. Some cats discovered his cache and managed to run away with 2 bones each as Arbax chased them away. Arbax is a smart dog who can count. He counted the bones and found that there were 17 left. How many cats were there?

Exercise 8

Lila has 18 new blue pens. Amira proposes an exchange: she will give Lila one fluorescent pen for every 7 blue pens. Lila wants 4 fluorescent pens. Amira will accept a scented eraser instead of 2 blue pens to make up for the difference. How many erasers does Lila have to give Amira?

Exercise 9

Which two consecutive months have 62 days in total?

Exercise 10

Five rats, two squirrels, and one mongoose decide to start a small business. In one day, each rat can bring in 2 clients and each squirrel can bring in 3 clients. The mongoose is in charge of the supply chain and has to provide 4 dollars worth of merchandise per client. How many days will it take the mongoose to spend 128 dollars?

Exercise 11

Each day, Penelope weaves 6 feet of fabric. Each night, she unravels 5 feet of fabric. How many days does it take her to weave a 12 ft long coverlet?

Exercise 12

An unusual weather pattern started on a Monday. Each day it rained for three hours as follows: the first day it rained from midnight to 3 am, the next day it rained from 3 am to 6 am, and so on. What day of the week was it when the rain started at midnight for the second time?

Exercise 13

Five bears foraged together in a forest. They decided to take turns to rest as follows: one of them would rest while the other four foraged. If each bear rested for 6 hours, how many hours did each bear spend foraging?

Exercise 14

It's dance recital day! Lila had her performance at 3 pm and arrived 2 hours ahead of time for rehearsal. Dina had her performance at 2 pm and arrived 3 hours ahead of time, when the hall opened, for rehearsal. Each performance was one hour long and the hall closed after Lila's performance. For how many hours was the hall open?

Exercise 15

Lila's train, which was supposed to arrive at 3:45 pm, arrived 20 minutes late. As a result, Lila missed the 3:50 bus and had to wait 25 minutes for the next bus. If the bus ride home takes 15 minutes, how late was Lila compared to her usual arrival time?

Exercise 16

Dina has a Book of Magic that is numbered backwards. If she is now on page 201 and the book has 453 pages, how many pages has she actually read?

Exercise 17

Cornelia, the shepherd, has 70 animals: sheep, hens, and cows. There are as many sheep's legs as hen's legs. There are as many cow's tails as hen's legs. How many cows are there?

Exercise 18

How many pounds of cabbage at 2 dollars a pound weigh the same as 5 pounds of onions at 1 dollar a pound?

Exercise 19

Lila had to help her mother mail some reports. She used a staple for every 3 sheets of paper. As she stapled, 20 staples broke and needed replacement. How many staples did she need if she had to staple 480 sheets in total?

Exercise 20

Ali and Baba are exploring the cave of the forty thieves. The entrance is locked, but Baba has a magic key that unlocks any lock. From the entrance, 5 corridors branch out, each of them locked. Each corridor has 3 rooms on one side and 3 rooms on the other side, all locked. How many of the rooms will Ali and Baba be able to explore if the magic key loses its power after 20 uses?

MISCELLANEOUS PRACTICE

Exercise 1

Jerome had twelve properly outfitted boats ready to be rented out at his beach sports shop. Each boat is required to have two paddles and four lifesavers. A storm swept away five paddles and eight lifesavers from the shop. How many operational boats did he have remaining?

Exercise 2

Lila is learning how to cook. She has a recipe for cookies that calls for 2 cups of flour, 3 sticks of butter, one cup of chocolate chips, and three eggs. She uses this recipe and makes 26 cookies. How many cups of flour should she use next time if the wants to bake 39 cookies?

Exercise 3

Dina, Lila, and Amira have 5 plush toys. Can you help them share the toys so that each of them has a different number of toys? (Each girl must have at least one toy.)

Exercise 4

Ali discovered a box filled with gold coins in the thieves' lair. Ali managed to stuff half of the coins in his pockets and Baba managed to stuff half of the remaining coins in his pockets before they heard noises from outside and slipped out of the cave. The returning thieves found 30 coins in the box. How many coins did Ali and Baba steal?

Exercise 5

If the letter **O** represents an odd digit, and the letter **E** represents an even digit, which of the following equalities is impossible?

(A) $O + O = E$

(B) $OO - OO = EE$

(C) $OO - O = EE$

(D) $EE + O = EO$

(E) $E + O = EO$

Exercise 6

Dina's grandmother has 5 blue cups, 2 red cups, 3 red saucers, 2 blue saucers, and 2 green saucers. How many cup and saucer pairs can she make for which the color of the saucer does not match the color of the cup?

Exercise 7

Tony, the car mechanic, wants to have equal amounts of coolant fluid in two containers. He has 4 liters of coolant in one container and 6 liters in the other. How many liters of coolant should he transfer from the first container into the second one?

Exercise 8

Dina and Amira are standing in line at the baker's shop. There are four people in front of Dina and four people between Dina and Amira. By the time Dina places her order, Amira will be:

(A) third in line

(B) fourth in line

(C) fifth in line

(D) sixth in line

(E) eighth in line

Exercise 9

Lila and Dina are playing "Little Romans." They have to dress in togas, speak Latin, and solve the following:

$$MCM - MC =$$

Exercise 10

Help Dina and Lila solve their Roman homework by providing results expressed as Roman numerals:

(a) $CM + MC =$

(b) $MCCC + DCCC =$

(c) $XI + IX =$

(d) $MC - CM =$

(e) $XI - IX =$

(f) $XXIV + XXVI =$

(g) $XXVI - XXIV =$

(h) $LX - XL =$

(i) $MMM - CCC =$

(j) $MCL - CML =$

(k) $CXX - LXX =$

(l) $CCC - XXX =$

Exercise 11

1. How many two digit numbers have identical digits?

2. How many three digit numbers have identical digits?

3. How many four digit numbers have identical digits?

4. How many one hundred digit numbers have identical digits?

Exercise 12

Lila has some cards with the digit 6 on them. By turning a card upside down, she can get a card with the digit 9 on it. Lila forms two digit numbers with these cards. What is the sum of all the different numbers she can make?

Exercise 13

In the figure, the square and the circle represent different digits. What is the difference between the largest and the smallest digit values the triangle can represent?

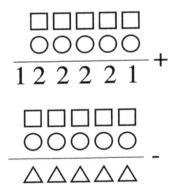

Exercise 14

The figure represents a correct addition:

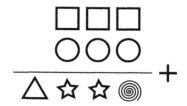

Different shapes represent different digits and the same shape always represents the same digit. Which of the following values can the spiral not have? Check all that apply.

(A) 0

(B) 1

(C) 7

(D) 8

(E) 9

Exercise 15

In the following list, how many numbers are odd?

$$4, \ 10, \ 151, \ 200, \ 0, \ 45, \ 736$$

Exercise 16

Find the following numbers:

    1. the largest three digit number with all different odd digits

    2. the smallest three digit number with all different odd digits

    3. the largest three digit number with all different even digits

    4. the smallest three digit number with all different even digits

Exercise 17

The following sequence is built according to a pattern. Help Dina add three more terms to it!

$$79, \ 69, \ 60, \ 52, \cdots$$

Exercise 18

The sum $\diamond + \diamond$, where $\diamond$ represents a digit, is:

(A) odd

(B) even

(C) impossible to determine

(D) neither even nor odd

Exercise 19

Amira has to make 3 different prizes for the swim meet. She has 5 toys to use as prizes, but each prize has to consist of a different number of toys. Can you explain why she is a bit puzzled?

Exercise 20

The largest of five consecutive odd numbers is 19. How many of these numbers are larger than 14?

Exercise 21

Which number does the star represent?

Exercise 22

Dina wrote four consecutive odd numbers, in increasing order, on a piece of paper. Arbax played with it and pierced it with his fangs. Now some of the digits are no longer legible. Can you figure out the missing digits?

$$\_\,3\,1\,\_$$
$$\_\,3\,\_\,9$$
$$5\,\_\,\_\,\_$$
$$\_\,\_\,\_\,\_$$

Exercise 23

Lila had 2 blue cubes worth 5 points each, 2 green cubes worth 3 points each, and 2 yellow cubes worth 6 points each. She placed them in a bag and asked Dina to pick 3 cubes without looking.

1. What is the largest odd total score Dina can obtain?
2. What is the lowest even total score Dina can obtain?

Exercise 24

Dina's mother organized a party for the children of her coworkers. She placed some boxes of candy on a table. Each box contained 20 candies. After the party, Lila noticed a funny thing had happened: one box was left untouched, from another only one candy had been removed, from another two candies were missing, from another three candies, and so on. The final box was completely empty. How many boxes were there in total?

Exercise 25

Lila and Amira are selling 40 of their old toys at a yard sale. Instead of selling each toy for 90 cents, they decided the sale would go quicker if they offered a "buy 3 toys and get one toy free" deal. To make the same amount of money in total, how much do they have to sell each toy for? (Assume they sell all the toys in either case.)

Exercise 26

Lila and Dina are trading toys. They agree that one doll should be worth 3 magnets and that one magnet should be worth two UV-beads. Lila has lots of UV-beads. If Lila wants to trade UV-beads for a doll and a magnet from Dina, how many UV-beads should she give Dina?

Exercise 27

Amira has stored 759 songs in her collection of favorites. How many more songs does she have to store for the number of songs to be the next number with three different odd digits?

Exercise 28

Dina and Lila are allowed to remove 3 digits from the number 13904521 but are not allowed to change the places of the remaining digits. Dina wants to obtain the largest possible number formed with the remaining digits, while Lila wants to obtain the smallest. What is the difference between Dina's number and Lila's number?

Exercise 29

Max, the baker, sells baguettes for 1.20 dollars each and focaccias for 3.40 dollars each. What is the smallest number of baguettes and foccacias he must sell in order to get an amount of money that can be written as whole dollars, no cents.

Exercise 30

For each puffin that flies off from a cliff, three puffins land on the cliff. There were 1000 puffins on the cliff and 20 of them flew off. How many puffins are there now on the cliff?

SOLUTIONS TO PRACTICE ONE

Exercise 1

Make a list of all the numbers from 1 to 30 that have only even digits.

Solution 1

2, 4, 6, 8, 20, 22, 24, 26, and 28

Exercise 2

What is the largest even 3-digit number?

Solution 2

998

Exercise 3

What is the largest 3-digit number with all digits even?

Solution 3

888

Exercise 4

A figure is painted on the side of a truck. If the truck has stopped beside a lake, how many digits can be seen in its reflection in the water?

Solution 4

Four of the ten digits: 0, 1, 3, and 8.

0123456789

0153ᵗᵉ↗8ᵨ

Exercise 5

How many different numbers can we make using the cards?

Solution 5

The card 86 turns into the card 98 when rotated half a circle.
The card 69, on the other hand, remains the same when rotated half a circle.

We can make 7 numbers:

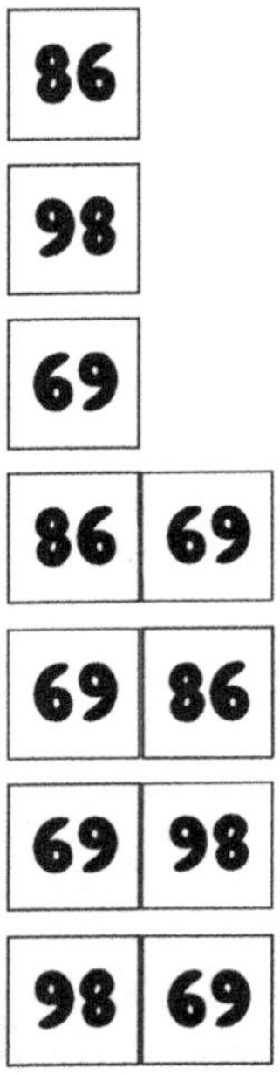

86

98

69

86	69

69	86

69	98

98	69

Solution 6

If we add two digits, the largest carryover we can get is 1 because the largest digit is 9.

Solution 7

If we add three digits, the largest carryover is 2, since $9 + 9 + 9 = 27$.

Solution 8

The answer is (B). Lila adds two digits and gets a carryover of 1. If one of the digits is smaller than 5, the other digit is necessarily larger than 5. Two digits that are both smaller than 5 do not produce a carryover when added.

Exercise 9

Dina tries to write the number 121212121212 as the sum of two numbers that only have digits of 1 and 0. Can you help her?

Solution 9

Two possible solutions are illustrated below:

$$
\begin{array}{r}
11111111111 \\
10101010101 \\
\hline
121212121212
\end{array} +
\qquad
\begin{array}{r}
110101010101 \\
1111111111 \\
\hline
121212121212
\end{array} +
$$

Solution 10

Both sums have the same value. Take, for example, the digit 1. It appears one time in the hundred thousands place, one time in the ten thousands, one time in the thousands, one time in the hundreds, one time in the tens, and one time in the units in *both* sums. The same is true about each of the other digits. Notice that we did not need to compute the actual values of the sums.

Exercise 11

If you count backwards from 999 to 1, at which count will the digit in the middle change for the first time?

Solution 11

At the 11th count:

1-st count	999
2-nd count	998
3-rd count	997
4-th count	996
5-th count	995
6-th count	994
7-th count	993
8-th count	992
9-th count	991
10-th count	990
11-th count	989

Exercise 12

How many three digit numbers can be formed using only the digits 7 and 0?

Solution 12

Strategic Approach: No number can start with 0. Therefore, we can only have numbers that start with 7. There is only one choice for the first digit. There are 2 choices for the second digit and 2 choices for the third digit. The total number of possibilities is $2 \times 2 = 4$.

Brute Force Approach: Make a list of the numbers:

$$\{700, 707, 770, 777\}$$

Exercise 13

Make a list of the digit sums for all the numbers between 11 and 20.

Solution 13

Strategic Approach: 11 has a digit sum of 2. As we count up from 11 to 19 only the last digit changes, each time by 1. So far, we will get all the sums between 2 and 10. When we reach 20 the digit sum will once again be 2. Therefore, the sums are all the consecutive integers from 2 to 10.

Brute Force Approach: Make a list of the digit sums:

11	2
12	3
13	4
14	5
15	6
16	7
17	8
18	9
19	10
20	2

Exercise 14

Lila made a list of all the numbers smaller than 100 that have a digit sum of 16. How many different digits did she use while writing the list?

Solution 14

16 is a pretty large sum for a two digit number. Only a few numbers can have such a large digit sum:

$$\{79,\ 88,\ 97\}$$

Lila used 3 different digits to write the numbers in her list.

Exercise 15

How many 5 digit numbers have a digit sum of 2?

Solution 15

There are 4 such numbers. Since no number can start with 0, there is only one number with a digit of 2 and 4 digits of zero: 20000. Then, there are numbers that have two digits of one and three digits of zero. These must all start with the digit 1; there are four other positions remaining for the other digit of 1:

$$\{10001,\ 10010,\ 10100,\ 11000\}$$

Exercise 16

Dina has written down a 6 digit number. Lila reverses the digits of this number. Then, the two girls compare their numbers and find that they are identical. What is the largest number of different digits Dina's number could have?

Solution 16

For the two numbers to be equal, Dina must have written down a *palindrome*. A 6-digit palindrome has at most three different digits. Example:

123321

186

Of course, it is possible to satisfy the condition with even fewer distinct digits, such as:

$$121121 \text{ or } 111111$$

Exercise 17

Lila asks Dina: "For how many even 3-digit numbers does the digit sum equal 26?" Dina finds a number. Can you help her find more?

Solution 17

No. Dina has found the only number that meets Lila's requirements: 998.

Exercise 18

The number 333 is written using only the digit 3. In how many ways can we write it as a sum of two positive numbers that are written using only the digits 3 and zero?

Solution 18

In three different ways:

$$
\begin{aligned}
333 &= 300 + 33 \\
333 &= 303 + 30 \\
333 &= 330 + 3
\end{aligned}
$$

Because addition is *commutative*, ways in which the same terms are added in a different order are not distinct from these.

Exercise 19

Dina made a list of all the numbers written by repeating the same digit that are greater than 405678 and smaller than 999110. Lila counted the numbers on Dina's list. How many numbers did Lila count?

Solution 19

Dina's list looks like this:

$$\{444444, 555555, 666666, 777777, 888888\}$$

Lila found that the list contains 5 numbers.

Exercise 20

Dina thinks of a number and Lila thinks of a number. When they add their numbers, the sum is even. If they subtract the smaller number from the larger number, is the difference even or odd?

Solution 20

If two integer numbers have an even sum, they also have an even difference.

Exercise 21

Lila could give Dina 5 beads in order for them to have the same number of beads. What is the difference between the number of beads Lila has and the number of beads Dina has?

Solution 21

Dina has 10 beads more than Lila:

Lila's beads		5	5

Dina's beads	

Solution 22

If the total number of beads can be divided into two equal parts, then it must be even. The impossible choice is 21 beads.

Exercise 23

What is the largest possible digit sum for a 3-digit number in which all the digits are different?

Solution 23

Choose the three largest digits: 9, 8, and 7. The largest digit sum is $7 + 8 + 9 = 24$.

Exercise 24

A computer begins counting by ones starting from 13579. What is the next number with odd digits that are all different?

Solution 24

13597

Exercise 25

Arbax, the Dalmatian, has 16 bones hidden in 5 different caches. Arbax thinks there is an odd number of bones in each cache. Is Arbax right?

Solution 25

Arbax cannot be right. If there is an odd number of bones in each cache and there is an odd number of caches, then the total number of bones must be odd. 16, however, is even.

$$O + O + O + O + O = E + E + O = E + O = O$$

Note that we do not need to assign specific numbers of bones to the caches.

SOLUTIONS TO PRACTICE TWO

Exercise 1

In the following figure, different shapes represent different digits, and the same shape represents the same digit. Which digit does the square represent?

Solution 1

The first number has two identical digits and is smaller than 42. Also, since we add only a single digit to obtain 42, the number must be equal to 33. The square represents the digit 9 ($42 - 33 = 9$).

Exercise 2

In the following sum, different letters represent different digits:

$$A + A + A + A + A = B$$

Find the digits!

Solution 2

Only $A = 1$ and $B = 5$ are possible. If $A = 0$ then $B = 0$ and this does not fulfill the requirement that they be different. Other values for A produce a carryover and result in a sum that is no longer a one-digit number.

Exercise 3

If A and B are digits and $A + A = B$, how many different values can A have?

Solution 3

Since the result of the addition is a single digit, there is no carryover. A cannot be larger than 4. Also, since **B** is a digit different from **A**, **A** cannot be zero. Therefore, **A** can be either 1, 2, 3, or 4. There are 4 possible values.

Exercise 4

Two different digits cannot have a difference of:

(A) 0

(B) 3

(C) 5

(D) 9

Solution 4

Two different digits cannot have a difference of 0. The other differences listed are possible.

Exercise 5

In the following figure, different shapes represent different digits and the same shape represents the same digit. Which digit does the square represent?

Solution 5

The square represents a single digit and cannot be greater than 9. The number formed of two circles, therefore, cannot be greater than

77. Since the circles represent the same digit, the only solution that satisfies the condition is:

$$77 - 9 = 68$$

The square represents the digit 9.

Exercise 6

Two numbers are encoded as AC and BC, where different letters represent different digits and the same letter always represents the same digit. Their sum could be any of the following, except:

(A) 100

(B) 102

(C) 103

(D) 104

Solution 6

Since the last digit of both numbers is the same, the sum must end with an even digit. The only answer choice that does not end with an even digit is (C).

Exercise 7

Each symbol represents a digit from 0 to 9. Different symbols represent different digits. Find out what the digits are. How many such additions are there?

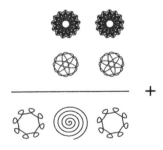

Solution 7

The two tens digits cannot add up to more than 18. Even if there is a carryover from the previous place value, the sum in the tens place cannot exceed 19. Therefore, the hundreds digit of the result must be 1. The units digit of the result has the same symbol, therefore it is also 1. Digits that add up to a number with the last digit 1 and that produce a carryover are: $5+6$, $4+7$, $3+8$, and $2+9$. Therefore, there are four possible solutions:

$$55 + 66 = 121$$
$$44 + 77 = 121$$
$$33 + 88 = 121$$
$$22 + 99 = 121$$

Exercise 8 Lila has found a solution to the following cryptarithm. Arbax, the Dalmatian, is also working hard to find a solution. Can he find a solution different from Lila's?

$$
\begin{array}{r}
\mathbf{XOX} \\
\mathbf{XOX} \\
\hline
\mathbf{OHH}
\end{array} \quad +
$$

Solution 8

When we add two identical digits the sum must be *even*. Therefore, **H** must represent an even digit. Since we have $\mathbf{X} + \mathbf{X} = \mathbf{H}$ in the units place and $\mathbf{X} + \mathbf{X} = \mathbf{O}$ in the hundreds place, one of these two sums must include a carryover. There can be no carryover in the units place, therefore we must have: $\mathbf{X} + \mathbf{X} + 1 = \mathbf{O}$ in the hundreds place. Hence, **O** must be odd.

Since the result ends in two identical digits (**HH**), they must come from adding the two different pairs of digits that produce the same last even digit.

- $0 + 0 = 0$ (no carryover)   or $5 + 5 = 10$ (carryover 1)

193

- $1 + 1 = 2$ (no carryover)   or $6 + 6 = 12$ (carryover 1)
- $2 + 2 = 4$ (no carryover)   or $7 + 7 = 14$ (carryover 1)
- $3 + 3 = 6$ (no carryover)   or $8 + 8 = 16$ (carryover 1)
- $4 + 4 = 8$ (no carryover)   or $9 + 9 = 18$ (carryover 1)

Since **O** is odd and **O** + **O** produces a carryover of 1, **O** can be either 5, 7, or 9. 5 is not possible because, in that case, **X** = 0 and 0 cannot be the first digit of a number. There remain two possible choices for **O**:

$$
\begin{array}{r}
272 \\
272 \\
\hline
544
\end{array} +
\qquad
\begin{array}{r}
494 \\
494 \\
\hline
988
\end{array} +
$$

Of these, only the one on the right satisfies the cryptarithm. Arbax cannot find a solution different from Lila's!

Exercise 9

Which digits do the symbols represent?

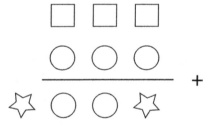

Solution 9

Since the sum of the square and the circle produce either a circle or a star, we conclude that the two digits produce a carryover when added. Since the carryover from adding two digits cannot exceed 1, the star must represent the digit 1 and the circle must be larger than the star by 1. Since the star is 1, the circle must be 2. 2 does not produce

194

a carryover when added to any digit except 9 or 8. In this case, the
square represents the digit 9. This is the solution:

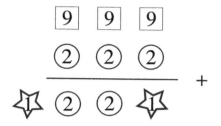

Exercise 10

Find the digit that corresponds to each letter:

$$
\begin{array}{cccc}
 & & & D \\
 & & D & D \\
 & D & D & D \\
\hline
A & A & B & C
\end{array} \; +
$$

Solution 10

A can only represent 1 because there cannot be a larger carryover from
the hundreds place. **D** must be a digit that, when added to some
carryover (which can be 1 or 2) produces a last digit of 1. Only 9
satisfies. Therefore, we must have:

$$
\begin{array}{cccc}
 & & & 9 \\
 & & 9 & 9 \\
 & 9 & 9 & 9 \\
\hline
1 & 1 & 0 & 7
\end{array} \; +
$$

Exercise 11

In the following examples, O represents an odd digit and E represents an even digit. Which example is impossible?

```
  O   E                    E   E
  E   O                    O   O
─────────  +  (A)      ─────────  +   (C)
  E   O                    O   O


  O   E                    O   E
  E   E                    O   O
─────────  +  (B)      ─────────  +   (D)
  O   O                    E   O
```

Solution 11

The impossible example is (B), because it shows two even units digits that produce an odd sum. Since there can be no carryover from a previous place value, this is not possible. In the other examples, a possible carryover of 1 from the units place to the tens place can change the parity of the digit accordingly. Here are some possible examples for each type of operation:

```
  3   8                    4   4
  2   5                    3   3
─────────  +  (A)      ─────────  +   (C)
  6   3                    7   7


  O   E                    3   4
  E   E                    3   5
─────────  +  (B)      ─────────  +   (D)
  O   O                    6   9
```

Exercise 12

In the following correct additions, the same symbol represents the same digit and different symbols represent different digits. Which sum is different from all others?

(A) ⬤◆ + ☐▽ =

(B) ◯▽ + ☐◆ =

(C) ▽☐ + ◯◆ =

(D) ☐◆ + ◯▽ =

(E) ☐▽ + ◯◆ =

Solution 12

Notice that, if the same digits are added in the same place values the sum is the same, even if the numbers are different. The choices (A), (B), (D), and (E) are all equivalent to:

$$10x\, \bigcirc + 10x\, \square + \triangledown + \blacklozenge =$$

whereas choice (C) is different.

Solutions to Practice Three

Exercise 1

Dina walked 6 steps up from the bottom of a staircase while Lila walked 5 steps down from the top. They met on the same step. How many steps does the staircase have?

Solution 1

The staircase has 11 steps.

Exercise 2

When Arbax, the Dalmatian, celebrates his 7th birthday, Lila will be 16. Today, it is Arbax's birthday and Lila is 10. How old is Arbax?

Solution 2

Arbax is one year old.

Exercise 3

Dina's mother is cooking stuffed peppers. For each pepper, she chops one tomato and two scallions. She fills a tray with 6 stuffed peppers. How many vegetables has she used in total?

Solution 3

She has used 6 tomatoes and 12 scallions. Don't forget that peppers are also vegetables and you have to add them to the total. The total number of vegetables she used is:

$$6 + 6 + 12 = 24$$

Exercise 4

There are four chairs at the breakfast counter in Lila's kitchen. At breakfast, Dina, Lila, and their parents sit down for their meal. How many legs can one count?

Solution 4

Four chairs have 16 legs. Four people have 8 legs. There are 24 legs in total.

Exercise 5

On a good spring day, one can see deer and hares from Dina's porch. Today, Dina saw as many hare's ears as deer's tails. If Dina saw 7 hares, how many deer did she see?

Solution 5

If Dina saw 7 hares, then she saw 14 hare's ears. Therefore, there were 14 deer's tails, which means Dina saw 14 deer.

Exercise 6

Lila gave Amira 12 magic wands. Amira gave Dina 8 magnets. Dina gave Lila 14 crystals. Which girl experienced the largest difference in her total number of toys? Was it an increase or a decrease?

Solution 6

Lila received 14 toys and gave away 12 toys. Her toys decreased in number by 2. Dina received 8 toys and gave away 14 toys. Her toys decreased in number by 6. Amira received 12 toys and gave away 8 toys. Her toys increased in number by 4. Dina's toys changed the most in number.

Exercise 7

Arbax, the Dalmatian, had stashed away 25 bones. Some cats discovered his cache and managed to run away with 2 bones each as Arbax chased them away. Arbax is a smart dog who can count. He counted the bones and found that there were 17 left. How many cats were there?

Solution 7

Arbax lost $25 - 17 = 8$ bones. If each cat stole 2 bones, there must have been 4 cats.

Exercise 8

Lila has 18 new blue pens. Amira proposes an exchange: she will give Lila one fluorescent pen for every 7 blue pens. Lila wants 4 fluorescent pens. Amira will accept a scented eraser instead of 2 blue pens to make up for the difference. How many erasers does Lila have to give Amira?

Solution 8

Lila can exchange 14 blue pens for 2 fluorescent pens and will have 4 blue pens left over. Lila needs another 14 blue pens to get 2 more fluorescent pens. She has 4 of these blue pens and is missing another 10, but she can pay Amira in scented erasers. Five scented erasers will make up for the 10 missing blue pens.

Exercise 9

Which two consecutive months have 62 days in total?

Solution 9

There are two solutions:

- July and August.
- December and January.

Exercise 10

Five rats, two squirrels, and one mongoose decide to start a small business. In one day, each rat can bring in 2 clients and each squirrel can bring in 3 clients. The mongoose is in charge of the supply chain and has to provide 4 dollars worth of merchandise per client. How many days will it take the mongoose to spend 128 dollars?

Solution 10

Each day, the rats bring in $5 \times 2 = 10$ clients, while the squirrels bring in $2 \times 3 = 6$ clients. The total number of clients is $10 + 6 = 16$ each day. If the mongoose spends 4 dollars for each client, the total daily amount he spends is:

$$16 + 16 + 16 + 16 = 64 \text{ dollars}$$

This happens to be the half of 128. It will take the mongoose only two days to spend 128 dollars!

Exercise 11

Each day, Penelope weaves 6 feet of fabric. Each night, she unravels 5 feet of fabric. How many days does it take her to weave a 12 ft long coverlet?

Solution 11

Within a 24-hour day, Penelope manages to add one foot of length to her coverlet. If this goes on for 6 days, the coverlet is then 6 feet long. On the seventh day, Penelope weaves 6 more feet to the length and... the coverlet is finished! She needs 7 days.

Exercise 12

An unusual weather pattern started on a Monday. Each day it rained for three hours as follows: the first day it rained from midnight to 3 am, the next day it rained from 3 am to 6 am, and so on. What day of the week was it when the rain started at midnight for the second time?

Solution 12

Draw a circle to represent a clock. Put a tick mark on it every three hours: at 12, 3, 6, and 9. On the outside, write down which day of the week it rained in that interval. After you get to noon, continue during the day until midnight - another 12 hours. This will take you to the next Monday. The rain will start again at midnight on a Tuesday.

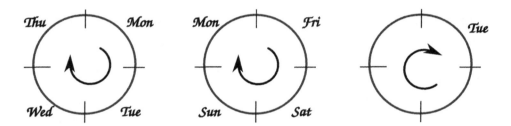

Exercise 13

Five bears foraged together in a forest. They decided to take turns to rest as follows: one of them would rest while the other four foraged. If each bear rested for 6 hours, how many hours did each bear spend foraging?

Solution 13

For each hour of rest, a bear foraged for 4 hours. For 6 hours of rest, the bear foraged for 24 hours.

Exercise 14

It's dance recital day! Lila had her performance at 3 pm and arrived 2 hours ahead of time for rehearsal. Dina had her performance at 2 pm and arrived 3 hours ahead of time, when the hall opened, for rehearsal. Each performance was one hour long and the hall closed after Lila's performance. For how many hours was the hall open?

Solution 14

The hall opened at 11 am, when Dina arrived. Lila performed from 3 pm to 4 pm, after which the hall closed. The hall was open for a total of 5 hours.

Exercise 15

Lila's train, which was supposed to arrive at 3:45 pm, arrived 20 minutes late. As a result, Lila missed the 3:50 bus and had to wait 25 minutes for the next bus. If the bus ride home takes 15 minutes, how late was Lila compared to her usual arrival time?

Solution 15

If the train had been on time, Lila would have taken the 3:50 bus.

However, the train arrived at 4:05 and Lila had to wait 25 minutes for the bus. Lila got on the bus home at 4:30 pm.

Since the bus ride home takes the same amount of time in both cases, it is not necessary to know how long it takes.

The difference in time between 3:50 pm and 4:30 pm is 40 minutes. Therefore, Lila arrived home 40 minutes later than usual.

Exercise 16

Dina has a Book of Magic that is numbered backwards. If she is now on page 201 and the book has 453 pages, how many pages has she actually read?

Solution 16

She has read $453 - 201 = 252$ pages.

Exercise 17

Cornelia, the shepherd, has 70 animals: sheep, hens, and cows. There are as many sheep's legs as hen's legs. There are as many cow's tails as hen's legs. How many cows are there?

Solution 17

Solve this problem by making a diagram. Since a sheep has four legs and a hen has only two legs, there must be more hens than sheep. Make a rectangle to represent the number of sheep:

 *Sheep*

The number of hens is going to be twice as large:

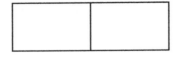 *Hens*

Since there are two cows for each pair of hen's legs, there are twice as many cows as hens.

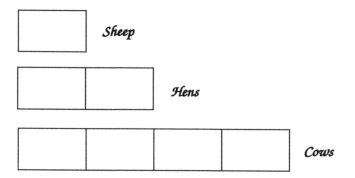

Count the small rectangles. There are 7 of them. If Cornelia has 70 animals, then each rectangle represents 10 animals. There must be 10 sheep, 20 hens, and 40 cows.

Exercise 18

How many pounds of cabbage at 2 dollars a pound weigh the same as 5 pounds of onions at 1 dollar a pound?

Solution 18

Since pounds are used to measures mass (weight), 5 pounds of cabbage will weigh the same as 5 pounds of onions. The other data does not play a role.

Exercise 19

Lila had to help her mother mail some reports. She used a staple for every 3 sheets of paper. As she stapled, 20 staples broke and needed replacement. How many staples did she need if she had to staple 480 sheets in total?

Solution 19

If Lila stapled 3 sheets together in a bundle, then she made 160 bundles out of the 480 sheets. She used 160 staples. Since 20 staples broke and needed replacement, the total number of staples she used was $160 + 20 = 180$.

Exercise 20

Ali and Baba are exploring the cave of the forty thieves. The entrance is locked, but Baba has a magic key that unlocks any lock. From the entrance, 5 corridors branch out, each of them locked. Each corridor has 3 rooms on one side and 3 rooms on the other side, all locked. How many of the rooms will Ali and Baba be able to explore if the magic key loses its power after 20 uses?

Solution 20

In the diagram, the circular shapes represent doors. Doors opened by using the magic key have been shaded:

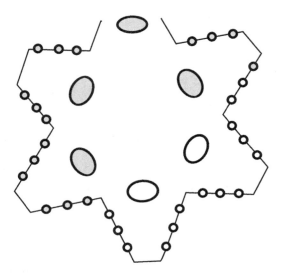

To explore all the rooms in a corridor, the key must be used 7 times. To open the main entrance and explore 2 corridors, Baba will use the key 15 times. Baba will be able to open another corridor and visit 4 of the 6 rooms in there. In total, they will be able to explore $6+6+4 = 16$ rooms.

SOLUTIONS TO MISCELLANEOUS PRACTICE

Exercise 1

Jerome had twelve properly outfitted boats ready to be rented out at his beach sports shop. Each boat is required to have two paddles and four lifesavers. A storm swept away five paddles and eight lifesavers from the shop. How many operational boats did he have remaining?

Solution 1

The loss of five paddles makes three boats unfit for use. The loss of eight lifesavers makes two boats unfit for use. Jerome will assign all the losses to as few boats as possible, leaving only three boats stranded on shore. Jerome will still be able to rent out 9 boats.

Exercise 2

Lila is learning how to cook. She has a recipe for cookies that calls for 2 cups of flour, 3 sticks of butter, one cup of chocolate chips, and three eggs. She uses this recipe and makes 26 cookies. How many cups of flour should she use next time if the wants to bake 39 cookies?

Solution 2

Each cup of flour makes 13 cookies. Three cups of flour will make 39 cookies.

Exercise 3

Dina, Lila, and Amira have 5 plush toys. Can you help them share the toys so that each of them has a different number of toys? (Each girl must have at least one toy.)

Solution 3

No, you cannot help them share the toys. If they each have a different number of toys, the smallest numbers they could have are 1, 2, and 3, for a total of 6 toys. Since they only have 5 toys, two of them will have the same number of toys, regardless of how they share them.

Exercise 4

Ali discovered a box filled with gold coins in the thieves' lair. Ali managed to stuff half of the coins in his pockets and Baba managed to stuff half of the remaining coins in his pockets, before they heard noises from outside and slipped out of the cave. The returning thieves found 30 coins in the box. How many coins did Ali and Baba steal?

Solution 4

Solve this problem by reasoning backwards from the end to the beginning of the story. If there were 30 coins left in the box and Baba had just taken half of the coins, then there were 60 coins in the box before Baba took his share. These 60 coins represented the half left by Ali, who must have taken the other 60 coins. Therefore, there were 120 coins in the box.

Exercise 5

If the letter **O** represents an odd digit, and the letter **E** represents an even digit, which equalities are impossible?

Solution 5

Start your solution to this problem by finding examples that work:

(A) $O + O = E$ $\qquad$ $3 + 5 = 8$

(B) $OO - OO = EE$ $\qquad$ $77 - 55 = 22$

(C) $OO - O = EE$ $\qquad$ $31 - 5 = 26$

(D) $EE + O = EO$ $\qquad$ $82 + 5 = 87$

(E) $E + O = EO$

The last equality is not possible, because the largest carryover from adding two digits is 1. Therefore, if there is a tens digit, it has to be equal to 1, which is not even.

Exercise 6

Dina's grandmother has 5 blue cups, 2 red cups, 3 red saucers, 2 blue saucers, and 2 green saucers. How many cup and saucer pairs can she make for which the color of the saucer does not match the color of the cup?

Solution 6

Represent the saucers by a string of letters:

RRRBBGG

Represent the cups by a string of letters:

BBBBBRR

Since the two green saucers do not match any of the cups, we can use the green saucers in any of the pairs. Assign the 2 blue saucers to the 2 red cups and the 3 red saucers to 3 of the blue cups. Finally, assign the green saucers to the remaining cups. You have obtained seven cup-saucer pairs:

BR, BR, BR, RB, RB, BG, BG

Exercise 7

Tony, the car mechanic, wants to have equal amounts of coolant fluid in two containers. He has 4 liters of coolant in one container and 6 liters in the other. How many liters of coolant should he transfer from the first container into the second one?

Solution 7

Tony has to pour 1 liter of fluid from the container with 6 liters into the other container. Now, there are 5 liters of coolant in each container.

Exercise 8

Dina and Amira are standing in line at the baker's shop. There are four people in front of Dina and four people between Dina and Amira. By the time Dina places her order, Amira will be:

(**A**) third in line

(**B**) fourth in line

(**C**) fifth in line

(**D**) sixth in line

(**E**) eighth in line

Solution 8

Dina is first in line when she is ordering. Count 4 more places down the line to reach Amira's position. Amira is 6th in line.

Exercise 9

Lila and Dina are playing "Little Romans." They have to dress in togas, speak Latin, and solve the following:

$$MCM - MC =$$

Solution 9

$$MCM - MC = DCCC$$

Solution 10

Provide results expressed as Roman numerals:

(a) $CM + MC = MM$

(b) $MCCC + DCCC = MMC$

(c) $XI + IX = XX$

(d) $MC - CM = CC$

(e) $XI - IX = II$

(f) $XXIV + XXVI = L$

(g) $XXVI - XXIV = II$

(h) $LX - XL = XX$

(i) $MMM - CCC = MMDCC$

(j) $MCL - CML = CC$

(k) $CXX - LXX = L$

(l) $CCC - XXX = CCLXX$

Exercise 11

1. How many two digit numbers have identical digits?

2. How many three digit numbers have identical digits?

3. How many four digit numbers have identical digits?

4. How many one hundred digit numbers have identical digits?

Solution 11

Remember that, if the number of digits of the number is specified and it is larger than 1, the number cannot start with a zero.

For each of the questions, there are 9 numbers with identical digits. Such a number is called a *repdigit*.

Exercise 12

Lila has some cards with the digit 6 on them. By turning a card upside down, she can get a card with the digit 9 on it. Lila forms two digit numbers with these cards. What is the sum of all the different numbers she can make?

Solution 12

$$66 + 69 + 96 + 99 = 330$$

Exercise 13

In the figure, the square and the circle represent different digits. What is the difference between the largest and the smallest digit values the triangle can represent?

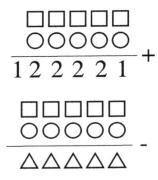

Solution 13

The sum of 122221 can be obtained only if there is a carryover from adding the square and the circle. Because the last digit of the number 12221 is 1, the square and the circle must add up to 11. Since the difference is positive, the square must represent a larger value than the circle. Possible pairs of digits that satisfy are: $6 + 5$, $7 + 4$, $8 + 3$, and $9 + 2$. The possible values of the digit represented by the triangle are: $6 - 5 = 1$, $7 - 4 = 3$, $8 - 3 = 5$, and $9 - 2 = 7$. The difference between the largest and smallest values is $7 - 1 = 6$.

Exercise 14

The figure represents a correct addition:

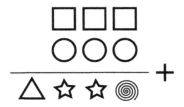

In the figure, different shapes represent different digits and the same shape always represents the same digit. Which of the following values can the spiral not have? Check all that apply.

Solution 14

The presence of the triangle means that the circle and the square must represent digits that add up to 10 or more, because there is a carryover. Therefore, we cannot achieve 9 in the place of the spiral by using additions such as: $1+8$, $2+7$, $3+6$, or $4+5$. Since two digits that produce a carryover add up to at most 18, the spiral cannot be 9. Also, because different symbols must represent different digits, the spiral cannot represent 0, 7, or 8.

Exercise 15

In the following list, how many numbers are odd?

$$\{4, 10, 151, 200, 0, 45, 736\}$$

Solution 15

4, 10, 200, 0, and 736 are even. 151 and 45 are odd. Therefore, two of the numbers are odd. **Attention!** 0 is even.

Solution 16

1. The largest three digit number with all different odd digits is 975.
2. The smallest three digit number with all different odd digits 135.
3. The largest three digit number with all different even digits 864.
4. The smallest three digit number with all different even digits 204.

Exercise 17

The following sequence is built according to a pattern. Help Dina add three more terms to it!

$$79, \ 69, \ 60, \ 52, \cdots$$

Solution 17

Notice that the difference between neighboring (*adjacent*) terms is a sequence of consecutive numbers:

$$
\begin{aligned}
79 - \mathbf{10} &= 69 \\
69 - \mathbf{9} &= 60 \\
60 - \mathbf{8} &= 52 \\
52 - \mathbf{7} &= 45 \\
44 - \mathbf{6} &= 39 \\
38 - \mathbf{5} &= 34
\end{aligned}
$$

Exercise 18

The sum $\diamond + \diamond$, where $\diamond$ represents a digit, is:

(A) odd

(B) even

(C) impossible to determine

(D) neither even nor odd

Solution 18

The result of the operation is twice the $\diamond$. It is, therefore, even. Moreover, for answer (D), note that no integer is neither even nor odd. All integers have well defined parity.

Exercise 19

Amira has to make 3 different prizes for the swim meet. She has 5 toys to use as prizes, but each prize has to consist of a different number of toys. Can you explain why she is a bit puzzled?

Solution 19

Even if Amira starts by giving only one toy to one of the prizewinners, she has to give a different number of toys to the next prizewinner. The smallest number of toys she can give the next winner is 2. Similarly, the smallest number of toys she can give the next winner is 3. Amira realizes that, to make three prizes with different numbers of toys, she must have at least 6 toys to distribute.

Exercise 20

The largest of five consecutive odd numbers is 19. How many of these numbers are greater than 14?

Solution 20

The five numbers are 11, 13, 15, 17, and 19. Three of these numbers are greater than 14.

Exercise 21

Which number does the star represent?

$$\bigcirc + \bigcirc + \bigcirc = 60$$

$$\text{(spiral)} + \text{(spiral)} = \bigcirc$$

$$\bigcirc + \text{(spiral)} = \text{(star)}$$

Solution 21

Since the three circles have a sum of 60, each circle must represent the number 20. Therefore, each spiral must represent 10 and the star must represent $20 + 10 = 30$.

Exercise 22

Dina wrote four consecutive odd numbers, in increasing order, on a piece of paper. Arbax played with it and pierced it with his fangs. Now some of the digits are no longer legible. Can you figure out the missing digits?

```
 _ 3 1 _

 _ 3 _ 9

 5 _ _ _

 _ _ _ _
```

Solution 22

Determine the last digit of the first number. It must be smaller than 9 by 2 - therefore, it is 7. Since the tens digit for the first number is 1, the hundreds and thousands digits will not change within this sequence. The numbers are: 5317, 5319, 5321, and 5323.

Exercise 23

Lila had 2 blue cubes worth 5 points each, 2 green cubes worth 3 points each, and 2 yellow cubes worth 6 points each. She placed them in a bag and asked Dina to pick 3 cubes without looking.

1. What is the largest odd total score Dina can obtain?
2. What is the lowest even total score Dina can obtain?

Solution 23

To obtain an odd score from 3 cubes, there must be 2 even scores and one odd score, or 3 odd scores. The largest odd score is 17:

$$\text{Odd} + \text{Even} + \text{Even} = \text{Odd}$$
$$5 + 6 + 6 = 17$$
$$\text{Odd} + \text{Odd} + \text{Odd} = \text{Odd}$$
$$5 + 5 + 3 = 13$$

To obtain an even score from 3 cubes, there must be 2 odd scores and one even score. It is not possible to obtain three even scores since there are only two cubes which are worth an even number of points.

$$\text{Even} + \text{Odd} + \text{Odd} = \text{Even}$$
$$6 + 3 + 3 = 12$$

The lowest even score is 12.

Exercise 24

Dina's mother organized a party for the children of her coworkers. She placed some boxes of candy on a table. Each box contained 20 candies. At the end of the party, Lila noticed a funny thing had happened: one box was left untouched, from another only one candy had been removed, from another two candies were missing, from another three candies, and so on. The final box was completely empty. How many boxes were there in total?

Solution 24

There were 21 boxes. 1, 2, 3, $\cdots$, 20 candies had been removed from different boxes. The last box, which was left empty, was the 20th box. Add the unopened box to these 20 boxes to find the total of 21.

Exercise 25

Lila and Amira are selling 40 of their old toys at a yard sale. Instead of selling each toy for 90 cents, they decided the sale would go quicker if they offered a "buy 3 toys and get one toy free" deal. To make the same amount of money in total, how much do they have to sell each toy for? (Assume they sell all the toys in either case.)

Solution 25

If they sell 4 toys for 90 cents each, they make 3 dollars and 60 cents. If they sell 3 toys for the same amount of money, they have to charge 1 dollar and 20 cents for each toy.

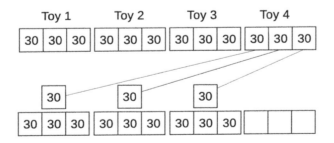

In the figure, we have represented the initial cost of each toy as a sum of the form $30 + 30 + 30 = 90$ cents. In order to be able to give one toy "for free," Lila and Amira charge an extra 30 cents for each of the three toys sold at regular price. The new price is $30 + 30 + 30 + 30 = 120$ cents (1 dollar 20 cents).

Exercise 26

Lila and Dina are trading toys. They agree that one doll should be worth 3 magnets and that one magnet should be worth two UV-beads. Lila has lots of UV-beads. If Lila wants to trade UV-beads for a doll and a magnet from Dina, how many UV-beads should she give Dina?

Solution 26

Use the diagram to find that one doll is worth 3 beads. Therefore, one magnet and one doll are worth $2 + 3 = 5$ beads.

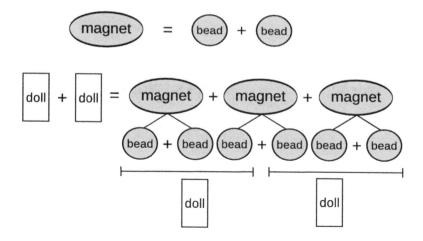

Exercise 27

Amira has stored 759 songs in her collection of favorites. How many more songs does she have to store for the number of songs to be the next number with three different odd digits?

Solution 27

The next number with different odd digits is 791. Amira must store another $791 - 759 = 32$ songs.

Exercise 28

Dina and Lila are allowed to remove 3 digits from the number 13904521 but are not allowed to change the places of the remaining digits. In this way, Dina wants to obtain the largest possible number formed with the remaining digits, while Lila wants to obtain the smallest. What is the difference between Dina's number and Lila's number?

Solution 28

Lila's work:	**Dina's work:**
1 3 9 0 4 5 2 1	1 3 9 0 4 5 2 1
$\cancel{1}\cancel{3}\cancel{9}$0 4 5 2 1	$\cancel{1}\cancel{3}$9$\cancel{0}$4 5 2 1
4 5 2 1	9 4 5 2 1

The difference is: $94,521 - 4,521 = 90,000$.

Exercise 29

Max, the baker, sells baguettes for 1.20 dollars each and focaccias for 2.30 dollars each. What is the smallest number of baguettes and foccacias he must sell in order to get an amount of money that can be written as whole dollars, no cents.

Solution 29

Two foccacias and one baguette will sell for $3.40 + 3.40 + 1.20 = 8$ dollars. Max has to sell three breads to obtain a whole number of dollars.

Exercise 30

For each puffin that flies off from a cliff, three birds land on the cliff. There were 1000 puffins on the cliff and 20 of them flew off. How many puffins are there now on the cliff?

Solution 30

Each time a bird flies off, the total number of birds on the cliff increases by 2. Since 20 birds flew off, the total number of birds on the cliff increased by 40. There are now 1040 puffins on the cliff.

PRACTICE OPERATIONS

OPERATIONS WITH INTEGERS

Notes to parents: The student has to understand the following aspects of operations:

1. The operation has a *syntax*, a way of writing it down so people and machines can understand what it means. This syntax can be changed. Civilizations have specified operations with numbers in different ways throughout history.

2. The *order of operations* is a construction that allows us to tell people and machines which operation should be the next in line to be performed. It is generally part of the *syntax* and it is just a convention. Students should observe several such conventions in order to get a feeling that syntax, as learned in school, is not cast in stone. We include a number of examples that use ad-hoc syntax which the student has to figure out.

3. While numbers are *abstract*, their *writing* is concrete. The symbols used to represent numbers, however, have changed through time. For example, the Arabic numeral 4 is different from the Roman numeral IV but both represent the same abstract notion. There is only a single number 4.

This book has several different goals:

- To help the student think flexibly, not mechanically, about operations.

- To help the student manipulate operations with numbers by using a variety of techniques in addition to the algorithms typically found in school textbooks.

- To develop a solid number sense by indicating what is changeable and what is not changeable in our arithmetic.

This section uses operations for both the first and the second grade level. Some students will be more ready, others less ready, for the entire content. If necessary, students can skip the more difficult problems at the first reading and come back to them a few months later.

Practice One

Do not use a calculator for any of the problems!

Exercise 1

Compute the sum:

$$16 + 0 =$$

Exercise 2

True or false?

$$102 - 0 = 102 + 50 - 50$$

Exercise 3

True or false?

$$102 + 0 = 102 - 50 + 50$$

Exercise 4

True or false?

$$5 + 2 - 2 = 2 - 2 + 5$$

227

Exercise 5

Dina computed the following:

$$100 + 11 - 11 + 12 - 12 + 13 - 13 =$$

$$100 - 11 + 11 - 12 + 12 - 13 + 13 =$$

She noticed something and found an explanation! She told Lila about it. What do you think Dina's explanation was?

Exercise 6

Lila computed the following very quickly:

$$100 + 99 + 98 + 97 + 96 + 95 + 94 - 95 - 96 - 97 - 98 - 99 - 100 =$$

How do you think she was able to do it so quickly?

Exercise 7

What is the value of the number covered by the clover?

$$77 - \clubsuit = 34$$

Exercise 8

What is the value of the number covered by the diamond?

$$108 - \diamondsuit = 99$$

Exercise 9

Dina performed these additions and noticed something interesting. What did Dina notice?

$$123 + 321 \ =$$

$$342 + 243 \ =$$

$$241 + 142 \ =$$

$$721 + 127 \ =$$

Exercise 10

Place the numbers 15, 12, and 11 in the squares to make the following equality true:

Exercise 11

Dina's calculator has a different way of understanding computations. To add two numbers, Dina has to enter + and then the two numbers she wants to add, like this:

$$+ \ 4 \ 5$$

When she presses **Enter**, the calculator displays the result: 9.
If she wants to subtract 4 from 5, Dina has to enter:

$$- \ 5 \ 4$$

The calculator scans the operations from left to right. It performs the operations that are directly followed by the numbers they apply to and replaces the whole group with the result. It then repeats this until only one result remains. For example:

If Dina wants to add 5 and 4 and then subtract 2 from the sum, which of the following is she going to have to enter?

(A) $+ \ 4 \ 5 - 2$

(B) $+ - \ 5 \ 4 \ 2$

(C) $- + \ 4 \ 5 \ 2$

Exercise 12

More practice with Dina's calculator:

(a) $+$ 7 8

(b) $+$ $+$ 7 8 2

(c) $-$ $+$ 7 8 2

(d) $+$ 5 3

(e) $-$ 5 3

(f) $\times$ 5 2

(g) $\div$ $\times$ 5 2 2

(h) $\div$ $+$ 5 5 2

(i) $\times$ $-$ 5 5 4

(j) $+$ $+$ $+$ 3 3 3 3

(k) $-$ $-$ $-$ 4 1 1 1

(l) $-$ $+$ 4 3 2

(m) $-$ $+$ 1 9 1

Note: This syntax is called the *Polish notation* and was invented by Jan Lukasiewicz in 1924.

Exercise 13

Fill in the missing values with non-zero one digit numbers.

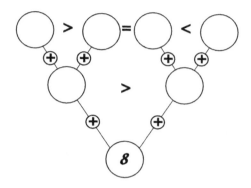

Exercise 14

By using parentheses, one can tell a calculator to perform the operation in the parentheses before the other operations. For example,

$$2 \times (5 - 2)$$

will instruct the calculator to do the subtraction first, then multiply the result by 2.

After performing the subtraction, the parentheses are no longer needed, and $(5 - 2)$ is replaced by 3, like this:

$$2 \times 3$$

Dina and Lila have entered the following operations in their calculators. Each calculator performs one operation at a time. Fill in the blanks with the numbers each calculator used for each step.

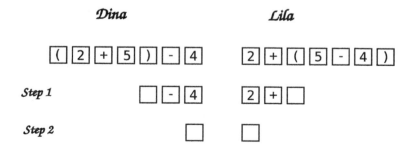

Exercise 15
Find a pattern and fill in the missing values:

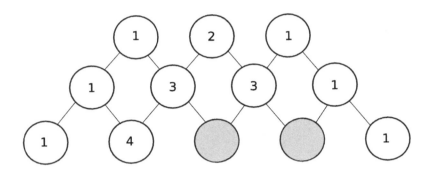

Exercise 16
Find the number in the grey circle:

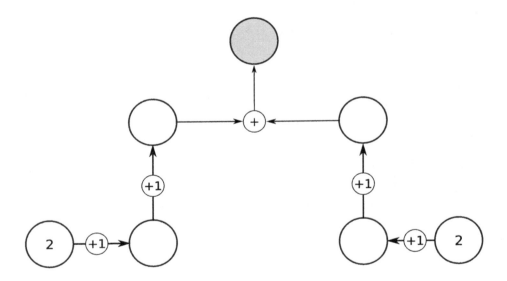

Exercise 17

Which of the following additions has an answer that is different from the others? How can you tell which it is without calculating all the sums?

(A) $11 + 22 + 33 + 44 =$

(B) $12 + 21 + 34 + 43 =$

(C) $14 + 23 + 31 + 42 =$

(D) $13 + 32 + 24 + 41 =$

(E) $41 + 32 + 24 + 42 =$

Exercise 18

Place the results of the operations in the boxes such that identical results go in the same box. How many boxes have remained empty?

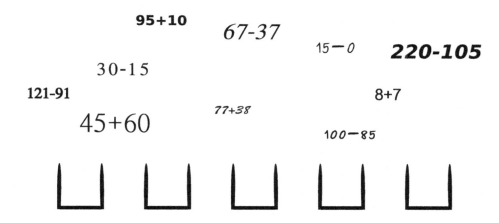

Using Negative Integers and Parentheses

A succession of additions and subtractions can be thought of as an **algebraic sum**: a sum of positive and negative numbers. By doing this, we can get rid of subtractions and one of the unpleasant properties associated with them: lack of commutativity.

Subtraction is not commutative:

$$7 - 4 \neq 4 - 7$$

However, addition is:

$$7 - 4 = -4 + 7$$

The advantage of introducing negative numbers is that they allow us to think of subtractions as additions. Now, the terms may be moved *provided we move them together with their sign:*

$$4 - 3 + 5 - 4 + 6 - 5 = 4 + 5 + 6 - 3 - 4 - 5 = 4 - 4 + 5 - 5 + 6 - 3$$

This distinction is often a source of confusion for students. They ask "How can you swap the terms? Didn't you say subtraction was not commutative?" Notice that we haven't swapped the terms around the minus sign! We have moved the term together with its sign:

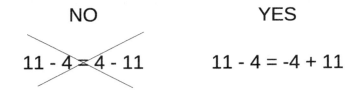

NO   YES

11 - 4 ≠ 4 - 11   11 - 4 = -4 + 11

234

Operations enclosed in parentheses have to be performed first. The parentheses are then replaced by the result (but not the operator in front of the leftmost parenthesis!)

If you are a machine:

1. Scan the expression *from left to right*.

2. Perform the operations in the *innermost* parentheses.

3. Replace the parentheses with the result.

4. Go back to step i. Stop when there is a single number left.

If you are *not* a machine, look at the expression carefully. Who knows, maybe there are some details that will help you find the result without doing so many computations!

Look at these two ways of processing operations:

$$3+4+5+6-2-3-4-5 \qquad 3+4+5+6-2-3-4-5$$
$$3+4+5+6-(2+3+4+5) \qquad 6-5+5-4+4-3+3-2$$
$$18-(14) \qquad 1+1+1+1$$
$$4 \qquad\qquad 4$$

Both of the procedures are better than performing each operation in turn from left to right. The procedure on the right hand side, however, is more efficient than the procedure on the left.

The procedure on the left illustrates that instead of subtracting a set of numbers, we can *subtract their sum*. Instead of working to remove parentheses, *we created them* because they helped us plan the order of our operations.

Practice Two

Exercise 1

True or false?

$$1 - 2 + 1 = 1 + 1 - 2$$

Exercise 2

True or false?

$$3 - 4 = 4 - 3$$

Exercise 3

True or false?

$$3 - 4 = -4 + 3$$

Exercise 4

Dina and Lila play a game on the number line. If Dina says a positive number, Lila will move that number of steps to the right. If Dina says a negative number, Lila will move that number of steps to the left. When Dina does not say anything, Lila stays put. Dina and Lila are facing each other. Dina says:

$$-2, 3, 1, -4, 2, -1, 3, -2, 1, 1, -4$$

Is Lila now to Dina's left or to Dina's right? How many steps away from Dina is Lila now?

Exercise 5

Lila must place the operators + and − in the empty squares so that the following equality is true. Can you help her?

Exercise 6

Compute efficiently:

(a) $22 - 5 + 1 - 8 =$

(b) $101 + 59 - 100 =$

(c) $119 + 1 - 49 - 1 =$

(d) $77 - 2 - 3 - 4 - 5 - 6 =$

(e) $299 + (1 - 299) =$

Exercise 7

True or false?

(a) $11 - 4 - 5 - 6 - 7 + 99 = 110 - (4 + 5 + 6 + 7)$

(b) $20 - 30 + 10 = 20 + 10 - 30$

(c) $2 - 3 + 4 - 5 + 6 = 6 - 5 + 4 - 3 + 2$

(d) $(13 + 12) - (11 + 14) = 12 - 11 + 13 - 14$

Exercise 8

True or false?

(a) $23 + 25 + 27 = 28 - 1 + 26 - 1 + 24 - 1$

(b) $16 + (1 - 16) = (16 - 16) + 1$

(c) $11 - (5 + 6) = 11 - 5 + 6$

(d) $5 + 15 + 25 - 20 - 10 = 5 + (15 - 10) + (25 - 20)$

Exercise 9

Lila has to place parentheses to obtain the smallest possible result:

$$43 \quad - \quad 3 \quad + \quad 15 \quad - \quad 7$$

Exercise 10

Dina has to place parentheses to obtain the result shown:

$$49 \quad - \quad 8 \quad - \quad 41 \quad + \quad 37 \quad = \quad 45$$

Exercise 11

Compute the result:

$$2 + 4 + 6 + 8 + 10 - 1 - 3 - 5 - 7 - 9 =$$

Exercise 12

Dina has to place parentheses in the following expression so that the computation is correct:

$$40 \quad + \quad 8 \quad \div \quad 8 \quad = \quad 6$$

Exercise 13

Lila has to place parentheses in the following expression so that the computation is correct:

$$9 \quad + \quad 8 \quad \times \quad 3 \quad = \quad 51$$

Exercise 14

Place parentheses in the following expression so that the computation is correct:

$$8 \ + \ 10 \ \times \ 2 \ = \ 36$$

Exercise 15

Place parentheses in the following expression so that the computation is correct:

$$4 \ + \ 5 \ \times \ 9 \ - \ 3 \ = \ 54$$

Exercise 16

Without calculating, can you tell whether each line is true or false?

$$10 - 1 + 11 - 1 \ = \ 10 + 11 - 2$$
$$8 - 3 \ = \ -3 + 8$$
$$8 - 3 \ = \ 3 - 8$$
$$5 + 3 - 3 \ = \ 8 + 5 - 8$$
$$11 - 12 + 12 - 1 \ = \ 11 - 1$$

Exercise 17

True or false?

$$2 - 2 + 3 - 3 + 4 - 4 + 5 - 5 = -2 + 2 - 3 + 3 - 4 + 4 - 4 + 5$$

Exercise 18

True or false?

$$5 + 6 + 7 + 8 + 9 - 5 - 6 - 7 - 8 - 9 = 5 + 6 + 7 + 8 + 9 - (5 + 6 + 7 + 8 + 9)$$

LARGE NUMBERS OF OPERANDS

At this age level, students will be able to count objects they see. For example, they will be able to count 6 numbers in this list: $\{1, 2, 3, 4, 5, 6\}$.

Let us imagine that the list is much longer, comprising the numbers from 1 to 40. We use the following notation to write the list:

$$1, \ 2, \ 3, \ 4, \ \ldots, \ 40$$

When asked how many numbers there are in the list, some students will answer 5 (which is the number of numbers they see) and others will answer 40 (which is the correct answer).

Students begin to understand that "there are more numbers behind the dots" at various ages, ranging from the elementary to the upper middle school grades. Precocious kids are able to work with this abstraction as early as first or second grade.

In this book, we try to provide exercises that familiarize the student with operating on large sets of numbers. Educators should take the student's current abilities into account when using this material and come back to it later if the student finds it too difficult.

Experiment

In the following experiment, we do not move or remove any of the cubes. We only hide some of them.

Look at this picture. How many cubes are there?

Look at this picture. How many cubes are there behind the transparent screen? How many cubes are there in total?

Look at this picture. How many cubes are there behind the opaque screen? How many cubes are there in total?

Look at this picture. How many cubes have been replaced by dots? How many cubes are there in total?

241

PRACTICE THREE

> Do not use a calculator for any of the problems!

Exercise 1

In the following sequence, dots have been used to replace some numbers that we do not want to write. Since the numbers follow a pattern, we can imagine the missing numbers without having to write them. Make a list of the missing numbers:

$$\{2, 4, 6, 8, 10, \ldots, 20\}$$

(a) How many numbers are written out?

(b) How many numbers have been replaced by dots?

(c) How many numbers are there in total (written and unwritten)?

Exercise 2

Make a list of the missing numbers:

$$\{2, 4, 6, 8, 10, \ldots, 40\}$$

(a) How many numbers are written out?

(b) How many numbers have been replaced by dots?

(c) How many numbers are there in total (written and unwritten)?

Exercise 3

Do not make a list of the missing numbers! There are too many of them. Based on your solutions to the previous two exercises, can you find out how many numbers have not been written out?

$$\{2, 4, 6, 8, 10, \ldots, 400\}$$

How many numbers are there in total in this sequence (written and unwritten)?

Exercise 4

Dina has to answer some questions about the following expression:

$$1 + 2 + 3 + \cdots + 40$$

1. How many numbers are added together?
2. How many of these numbers are odd?
3. How many + operators are there in total?
4. Will the result be even or odd? (Answer without calculating the result.)

Exercise 5

How many squares make a triangle?

$$\square + \square + \square + \square = \triangle + \square$$

Exercise 6

How many squares make a star?

$$\underbrace{\square + \square + \cdots + \square}_{101 \text{ squares}} = \star + \square$$

Exercise 7

How many squares make a circle?

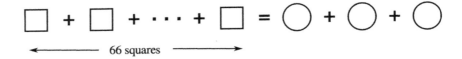

Exercise 8

Dina has to count the number of operations in the following sum:

$$1 + 2 + 3 + 4 + 5 + 6 + 7 + 8 + 9 + 10$$

What has Dina found?

Exercise 9

Lila has to count the number of operations in the following sum:

$$1 + 2 + 3 + 4 + 5 + \cdots + 100$$

If Lila knows what Dina found in the previous problem, can she find an answer without writing out all the numbers?

Exercise 10

Compute efficiently:

$$2 + 4 + 6 + 8 + 10 + \cdots + 50 - 1 - 3 - 5 - 7 - 9 - \cdots - 49 =$$

Exercise 11

How many numbers are there in the list?

$$\{11,\ 22, \cdots, 99,\ 111,\ 222, \cdots, 999\}$$

Exercise 12

How many numbers are there in the list?

$$\{1,\ 11,\ 111,\ 1111, \cdots, 111111111111111\}$$

Exercise 13

How many numbers in the list are even?

$$\{0,\ 1,\ 2,\ 3, \cdots, 100\}$$

Exercise 14

How many towers are there?

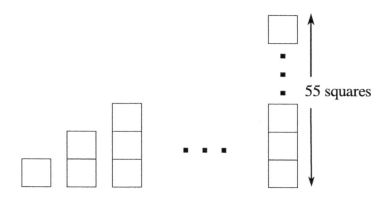

Exercise 15

How many circles are shaded?

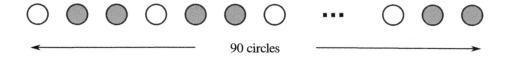

Exercise 16

How many digits have been used to write the following list:

$$\{1,\ 22,\ 333, \cdots, 7777777, \cdots, 333,\ 22,\ 1\}$$

Exercise 17

Lila and Dina are playing a game of dots. They use a ruled sheet of paper. Dina starts by drawing a dot on the first line. Lila draws two dots on the second line. Dina draws three dots on the third line. They continue on until one of them draws 47 dots on a line. Which line is it? Who draws the 47 dots, Lila or Dina?

Exercise 18

Complete the missing numbers in the exercises, according to the model:

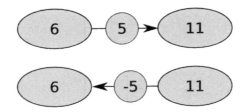

Pay attention to the direction of the arrow!

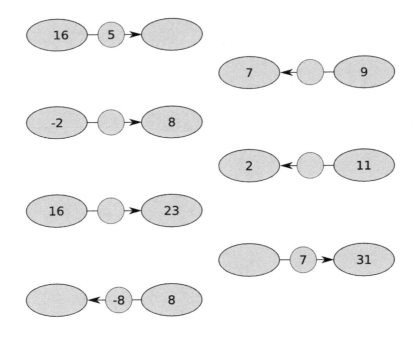

Understanding Equations

It is not too early to familiarize the student with equations.
First of all, start with the obvious observation:

$$4 = 4$$

Now let us say the square represents some number. Also, the square *always* represents the same number. The fact that the square never changes its value, enables us to say that:

$$\square \;=\; \square$$

The equality is an *identity* - it states that the square on the left is identical to the square on the right. This will be true if we choose the square to be the number 5 and it will also be true if we choose the square to be the letter Q.

The following, however, is not an identity anymore:

$$\square \;+\; \bigcirc \;=\; \square \;+\; \mathbf{3}$$

The square can still be anything but the circle must be the number 3.

This last situation is called an *equation*. It is an equality that is true *only for certain values* of the symbols used.

Problems with sum and difference of two numbers can be solved using a model with boxes.

Use a box for one of the numbers and a different size box for the other number (just in case they are not equal), like this:

Now draw the sum and the difference:

SUM

DIFFERENCE

Notice that, if you subtract the difference from the sum, you get twice as much as the smaller number:

Twice the smaller number.

From here, you can find both numbers.

Example 1: Tom has 5 absences from choir practice more than Evelyn. Together, they have missed choir practice 11 times this year. How many absences does each have?

Solution: The sum of their absences is 11 and the difference is 5. Subtract the difference from the sum: $11 - 5 = 6$. The smaller number is the half of 6. Therefore, Evelyn has 3 absences and Tom has $3 + 5 = 8$.

Always draw a model with boxes and make sure you understand the diagram before doing computations. Even better, **build** a model with Lego blocks!

Example 2: In 2015, there were 71 more rainy days than sunny days in Weatherville. How many sunny days were there?

Solution: Since 2015 cannot be a leap year, it must have had 365 days in total (sum), of which there were 71 (difference) more rainy days than sunny days.

The number of sunny days is smaller than the number of rainy days.

Subtract the difference from the sum:

$$365 - 71 = 294$$

This number represents twice the number of sunny days. If you do not know how to divide, make two equal groups in the way that seems easiest to you. I would do:

$$
\begin{aligned}
200 &= 100 + 100 \\
80 &= 40 + 40 \\
14 &= 7 = 7
\end{aligned}
$$

There have been 147 sunny days.

PRACTICE FOUR

Do not use a calculator for any of the problems!

Exercise 1

Which number is hiding behind the square?

$$\square + \square = \textbf{40}$$

Exercise 2

Which number is hiding behind the circle?

$$\bigcirc + \bigcirc = \textbf{22}$$

Exercise 3

Fill in the circles with appropriate non-zero positive integers:

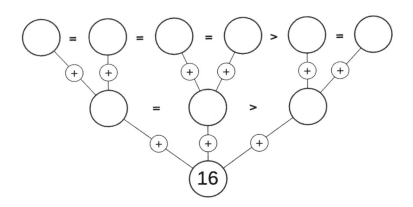

Exercise 4

Find K:

$$61 + 59 = K + K + K$$

Exercise 5

Place the same number in each empty circle ('/' means 'divide'). Which number must it be?

Exercise 6

To make the equality true, place in the circles numbers formed using the same digit:

Exercise 7

Place operators $(+, -, \times, \div)$ within the boxes. No parentheses are needed.

Exercise 8

In each of the following, which number does the letter represent?

$$A + A + A = 15$$

$$B + B + B + B - B = 15$$

$$C + C + C + C + C = 15$$

$$C + C + C + C + C + C - C = 15$$

$$D + D - D + D - D + D - D + D - D + D - D = 15$$

Exercise 9

The triangle, the circle, and the square have different weights. Which of the following would balance the scale on the right?

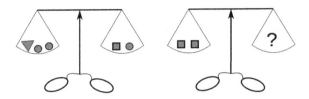

(A) one triangle and one circle

(B) one triangle and three circles

(C) two triangles and one circle

(D) two triangles and two circles

Exercise 10

Amira is out shopping for Hallowe'en treats. She wants to buy some Real Fruit Eyeballs and some Fruit Jelly Smellies. Three packs of Eyeballs and two packs of Smellies cost 9 dollars. Two packs of Eyeballs and three packs of Smellies cost 11 dollars. Amira has 12 dollars. Does she have enough money to buy three packs of Eyeballs and three packs of Smellies?

Exercise 11

Dina and Lila have traveled to their grandparents' home a total of 30 times. Of these, Lila traveled on 6 fewer occasions than Dina. How many times did each of them visit their grandparents?

Exercise 12

A number is smaller than another by 45. The sum of the two numbers is 99. Find the two numbers.

Exercise 13

Arbax has given Lynda 12 of his treats. Now, Lynda has 5 treats more than Arbax. How many treats more than Lynda did Arbax have?

Exercise 14

Amira has lost 14 of her action figures. Dina has lost only 10 of her action figures. Lila tells them: 'We used to have 104 action figures together. Now that you lost some of yours, I have as many action figures as both of you.' How many action figures does Lila have now?

Exercise 15

Lila, Dina, and Amira are picking flowers to decorate the living room. Lila picks 5 more flowers than Dina. Amira picks 3 flowers less than Dina. The whole bunch is made up of 23 flowers. How many flowers did each of them pick?

MISCELLANEOUS PRACTICE

Do not use a calculator for any of the problems!

Exercise 1

Which number does the circle represent?

$$\bigcirc + \bigcirc = \star$$

$$\star + \star = 136$$

Exercise 2

Replace the question mark with a number so that the operations are correct:

$$\bigcirc + \star = 16$$

$$\bigcirc - \star = 4$$

$$\bigcirc + \bigcirc = ?$$

Exercise 3

If two positive integers have an odd difference, is their sum:

(A) always even?

(B) always odd?

(C) sometimes even and sometimes odd?

Exercise 4

There are 100 plus signs in the following operation. What is the result of the additions?

$$1+1+1+\cdots+1= ?$$

Exercise 5

Roman numerals time! Perform the following operations and write the answer in both Roman and Arabic numerals:

(a) I + I =

(b) II + II =

(c) III + III =

(d) IV + IV =

(e) V + V =

(f) XX + XX =

(g) XXX + XXX =

(h) LX + LX =

(i) L + L =

(j) C + C =

(k) CC + CC =

(l) CCC + CCC =

(m) CD + CD =

(n) CD + D =

Exercise 6

Each group of four numbers in the sequence follow the same pattern. Which numbers belong in the empty circles?

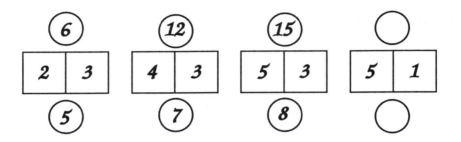

Exercise 7

In the figure, shaded circles of the same color hide the same number. Which number does the white circle hide?

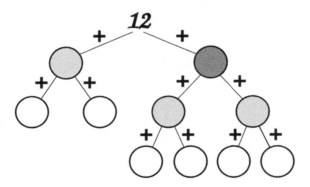

Exercise 8

Compute:

$$4 + 3 - 4 + 5 - 6 + 6 - 7 + 8 - 9 =$$

Exercise 9

Dina had 30 party balloons. During the party, 11 balloons popped. Afterwards, Lila gave Dina as many balloons as had popped. How many party balloons did Dina have then?

Exercise 10

The table in the figure has 10 rows and 11 columns. How many grey cells are there in the table? How many grey cells should we color white in order to have an equal number of white and grey cells?

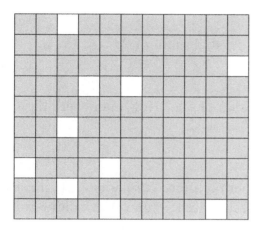

Exercise 11

In the lists below, some digits, not necessarily identical, have been hidden by a ♠. In each list, there are two different 3-digit numbers. Find the lists in which it is possible to tell which number is smaller.

(A) 9♠7, ♠87

(B) 33♠, 31♠

(C) 4♠3, 5♠♠

(D) 9♠♠, 9♠0

Exercise 12

Dina has 20 magnets and Lila has 11 magnets. How many magnets could Amira have if she has fewer than Dina and more than Lila? Check all that apply.

(A) 9

(B) 10

(C) 13

(D) 19

Exercise 13

How many 3-digit numbers have a digit sum of 2?

Exercise 14

A 100-digit number has a digit sum of one. How many zero digits does the number have?

Exercise 15

Dina has written a 3 digit number on a piece of paper. Lila must guess which number it is. Dina gives Lila a hint: "One of the digits is 5. The number does not change if you move the last digit in front of the first." Which number is it?

Exercise 16

Compute:

(a) $10 - 9 + 8 - 7 + 6 - 5 + 4 - 3 + 2 - 1 + 0 =$

(b) $100 + 99 - 99 + 98 - 98 + 97 - 97 + 96 - 96 + 95 - 95 + 94 - 94 + 93 - 93 + 92 - 92 + 91 - 91 + 1 =$

(c) $9 - 2 + 3 - 4 + 5 =$

Exercise 17

Lila has 5 more toys than Amira does. If Lila gives Amira 7 toys, how many more toys than Lila will Amira have?

Exercise 18

Find the positive integers that are hidden by symbols. Different symbols hide different integers. How many different solutions can be found?

$$\Diamond + \heartsuit + \heartsuit = 7$$

Exercise 19

In the following sequence of consecutive odd numbers, how many numbers have been replaced by dots?

$$15, \ 17, \ \cdots, \ 25$$

Exercise 20

How many numbers between 200 and 240 can be written using only the digits 2 and 3?

Exercise 21

A machine crunches numbers and outputs the result. The figure shows a set of numbers entering the machine and the results of the crunches coming out of the machine. If the machine is given the number 207 to crunch, what will it output?

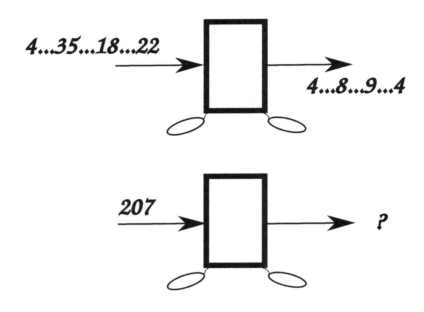

Exercise 22

Another machine mashes numbers and outputs the result. The figure shows a set of numbers entering the machine and the results of the mashes coming out of the machine. If the machine is given the number 207 to mash, what will it output?

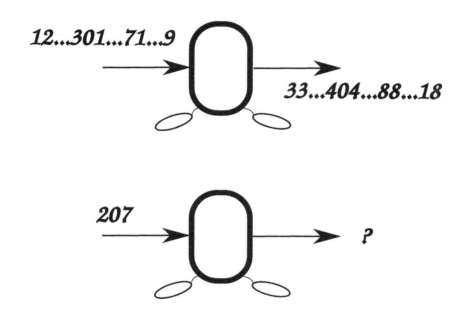

SOLUTIONS TO PRACTICE ONE

Solution 1

$$16 + 0 = 16$$

Exercise 2

True or false?
$$102 - 0 = 102 + 50 - 50$$

Solution 2

Since $50 - 50$ equals zero, the left hand side is equal to the right hand side. True.

Exercise 3

True or false?
$$102 + 0 = 102 - 50 + 50$$

Solution 3

This exercise is similar to the one above, except we now subtract 50 *before* adding it back in. Subtracting and then adding the same quantity from the original number does not change it, so the result is still 102. True.

Exercise 4

True or false?

$$5 + 2 - 2 = 2 - 2 + 5$$

Solution 4

True. Adding and then subtracting 2 from a number leaves the number unchanged.

The right hand side models the following situation: "Dina places two one dollar bills on the table. She then removes them and places a five dollar bill on the table. How many dollars are there on the table now?"

The left hand side models the following situation: "Dina places a five dollar bill on the table. She then places two one dollar bills on the table. She then removes the two one dollar bills. How many dollars are there on the table now?"

Exercise 5

Dina computed the following:

$$100 + 11 - 11 + 12 - 12 + 13 - 13 =$$

$$100 - 11 + 11 - 12 + 12 - 13 + 13 =$$

Solution 5

Dina noticed that she gets the same result:

$$100 + 11 - 11 + 12 - 12 + 13 - 13 = 100$$

$$100 - 11 + 11 - 12 + 12 - 13 + 13 = 100$$

In both cases, we add and then subtract an 11. The same is true for 12 and 13 and, therefore, the initial number (100) remains unchanged.

Dina said: "If you add and subtract the same number from another, the initial number does not change. It does not make a difference if you add first and then subtract or the other way around."

Lila said: "Wow! Then we don't actually need to do any computations. We just have to notice how many add-subtract pairs of the same number there are and cross them out!"

Think creatively, do not just perform all operations from left to right.

Exercise 6

$$100 + 99 + 98 + 97 + 96 + 95 + 94 - 95 - 96 - 97 - 98 - 99 - 100 =$$

Solution 6

Lila noticed that there were a lot of add-subtract pairs:

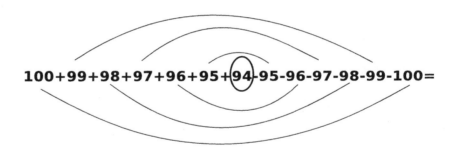

She crossed out the instances where the same number was both added and subtracted and was left with the number 94.

Exercise 7

What is the value of the number covered by the clover?

$$77 - \clubsuit = 34$$

Solution 7

43

Exercise 8

What is the value of the number covered by the diamond?

$$108 - \diamondsuit = 99$$

Solution 8

9

Exercise 9

Dina performed these additions and noticed something interesting. What did Dina notice?

$$123 + 321 = 444$$
$$342 + 243 = 545$$
$$241 + 142 = 383$$
$$721 + 127 = 848$$

Solution 9

Dina noticed that each of the results is a palindrome (a number that remains unchanged when the order of its digits is reversed).

Exercise 10

Place the numbers 15, 12, and 11 in the squares to make the following equality true:

$$\square \quad + \quad \square \quad - \quad \square \quad = \quad \mathbf{16}$$

Solution 10

A possible solution is:

$$\boxed{12} \quad + \quad \boxed{15} \quad - \quad \boxed{11} \quad = \quad 16$$

Another solution is possible if we use the commutative property of addition.

Exercise 11

If Dina wants to add 5 and 4 and then subtract 2 from the sum, what is she going to have to enter?

(A) $+$ 4 5 $-$ 2

(B) $+$ $-$ 5 4 2

(C) $-$ $+$ 4 5 2

Solution 11

If Dina enters choice (A), the calculator will complain like this: "SYN-TAX ERROR." This means that it does not understand the input. While $+$ 4 5 produces the answer 9, the calculator does not know what to do with the 9 $-$ 2 input since it expects the operator to be *in front of* the numbers it applies to.

If Dina enters choice (B), the calculator will find it knows how to compute $-$ 5 4. It will use the result (1) in the next operation. After this, the calculator will see the input $+$ 1 5. It can understand this, since the operator is in front of the two numbers. It will produce the answer 6.

If Dina enters choice (C), the calculator will find it knows how to compute + 4 5 and it will get 9. After this is done, it will see the new input, − 9 2, which it knows how to compute. It will get the answer 7.

The correct answer is (C).

Solution 12

The calculator reads the input from left to right and immediately performs the operations it understands. It replaces the operation with its result and reads the input again from left to right, like in this diagram:

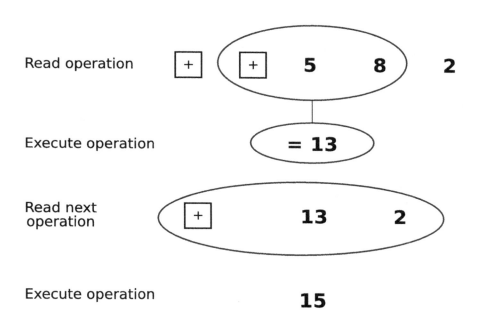

(a) + 7 8 = 15

(b) + + 7 8 2 = + 15 2 = 17

(c) − + 7 8 2 = − 15 2 = 13

(d) $+$ 5 3 $=$ 8

(e) $-$ 5 3 $=$ 2

(f) $\times$ 5 2 $=$ 10

(g) $\div$ $\times$ 5 2 2 $=$ $\div$ 10 2 $=$ 5

(h) $\div$ $+$ 5 5 2 $=$ $\div$ 10 2 $=$ 5

(i) $\times$ $-$ 5 5 4 $=$ $\times$ 0 4 $=$ 0

(j) $+$ $+$ $+$ 3 3 3 3 $=$ $+$ $+$ 6 3 3 $=$ $+$ 9 3 $=$ 12

(k) $-$ $-$ $-$ 4 1 1 1 $=$ $-$ $-$ 3 1 1 $=$ $-$ 2 1 $=$ 1

(l) $-$ $+$ 4 3 2 $=$ $-$ 7 2 $=$ 5

(m) $-$ $+$ 1 9 1 $=$ $-$ 10 1 $=$ 9

Exercise 13

Fill in the missing values with one digit non-zero numbers.

Solution 13

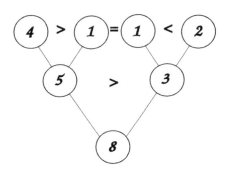

Exercise 14

Dina and Lila have entered the following operations in their calculators. Each calculator performs one operation at a time. Fill in the blanks with the numbers each calculator used for each step.

Solution 14

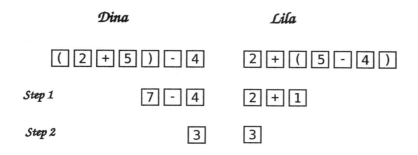

Dina and Lila have found out that: if the only operations are addition and subtraction, it is not necessary to perform them left to right. We can perform either one first, then the other, and obtain the same result.

Exercise 15

Find a pattern and fill in the missing values.

Solution 15

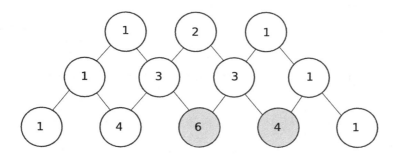

Exercise 16

Find the number in the grey circle.

Solution 16

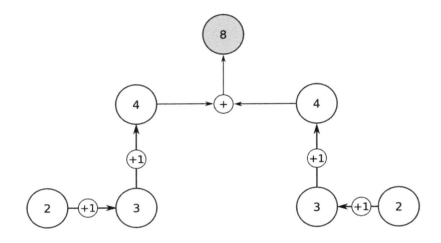

Exercise 17

Which of the following additions has an answer that is different from the others? How can you tell which it is without calculating all the sums?

(**A**) $11 + 22 + 33 + 44 =$

(**B**) $12 + 21 + 34 + 43 =$

(**C**) $14 + 23 + 31 + 42 =$

(**D**) $13 + 32 + 24 + 41 =$

(**E**) $41 + 32 + 24 + 42 =$

Solution 17

Without calculating, notice which sums have the same digits in the same place values. For example:

$$13 + 32 + 24 + 41 = 10 + 3 + 30 + 2 + 20 + 4 + 40 + 1$$

If you look at all the terms as a certain number of tens plus a certain number of ones, you can see that in all exercises except for (**E**), there are the same tens and the same ones added up.

Exercise 18

Place the results of the operations in the boxes such that identical results go in the same box. How many boxes have remained empty?

Solution 18

The results are:

$$95 + 10 = 105$$
$$15 - 0 = 15$$
$$100 - 85 = 15$$
$$220 - 105 = 115$$
$$77 + 38 = 115$$
$$8 + 7 = 15$$
$$45 + 60 = 105$$
$$30 - 15 = 15$$
$$121 - 91 = 30$$
$$67 - 37 = 30$$

There are only 4 distinct results. Therefore, one box remains empty.

SOLUTIONS TO PRACTICE TWO

Exercise 1

True or false?

$$1 - 2 + 1 = 1 + 1 - 2$$

Solution 1

On both sides, we add 1 twice and subtract 2. The result is zero on both sides.

True.

Exercise 2

True or false?

$$3 - 4 = 4 - 3$$

Solution 2

On the left hand side, we add 3 and subtract 4: the result is -1.
On the right hand side, we add 4 and subtract 3: the result is 1.
False.

Exercise 3

True or false?

$$3 - 4 = -4 + 3$$

Solution 3

On the left hand side, we add 3 and subtract 4: the result is -1.
On the right hand side, we add 3 and subtract 4: the result is -1.
True.

Exercise 4

Dina and Lila play a game on the number line. If Dina says a positive number, Lila will move that number of steps to the right. If Dina says a negative number, Lila will move that number of steps to the left. When Dina does not say anything, Lila stays put. Dina and Lila are facing each other. Dina says:

$$-2, \ 3, \ 1, \ -4, \ 2, \ -1, \ 3, \ -2, \ 1, \ 1, \ -4$$

Is Lila now to Dina's left or to Dina's right? How many steps away from Dina is Lila now?

Solution 4

Make a model of a number line and move the tip of the pencil left and right from the starting position.

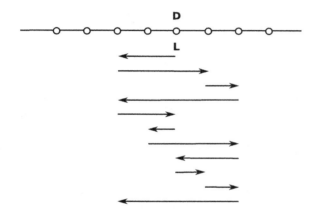

Lila ends up two steps away from Dina's right.
This result can also be obtained by performing the operations:

$$-2 + 3 + 1 - 4 + 2 - 1 + 3 - 2 + 1 + 1 - 4 = -2$$

Exercise 5

Lila must place the operators $+$ and $-$ in the empty squares so that the following equality is true. Can you help her?

Solution 5

Exercise 6

Compute efficiently:

Solution 6

(a) $22 - 5 + 1 - 8 = 10$

Add the positive numbers together and the negative numbers together: $22 + 1 = 23$ and $5 + 8 = 13$. Now, the operation is:

$$23 - 13 = 10$$

(b) $101 + 59 - 100 = 60$

Subtract 100 from 101 and get 1. Add 1 to 59 to get 60.

(c) $119 + 1 - 49 - 1 = 70$

Cross out the $+1$ -1 pair. You are left with:

$$119 - 49 = 70$$

(d) $77 - 2 - 3 - 4 - 5 - 6 = 57$

Add together all the numbers that have to be subtracted from 77. Add 4 and 6 to get 10. Add 2, 3, and 5 to get 10. All the numbers to be subtracted add up to 20. Subtract 20 from 77 to get 57.

(e) $299 + 1 - 299 = 1$

Notice that you are adding and then subtracting 299. Cross out both terms. You are left with 1.

Exercise 7

True or false?

Solution 7

(a) $11 - 4 - 5 - 6 - 7 + 99 = 110 - (4 + 5 + 6 + 7)$

Add together the positive values on both sides of the equation, then add together the negative numbers. On both sides, 11 and 99 are added and 4, 5, 6, and 7 are subtracted. The equality is true.

(b) $20 - 30 + 10 = 20 + 10 - 30$

True.

(c) $2 - 3 + 4 - 5 + 6 = 6 - 5 + 4 - 3 + 2$

On both sides, 2, 4, and 6 are added and 3 and 5 are subtracted. Both sides produce the same result. True.

(d) $(13 + 12) - (11 + 14) = 12 - 11 + 13 - 14$

Notice how, on the right hand side, 12 and 13 are added and 11 and 14 are subtracted. This is the same as subtracting the sum of 11 and 14 from the sum of 13 and 12. The equality is true.

Exercise 8

True or false?

Solution 8

(a) $23 + 25 + 27 = 28 - 1 + 26 - 1 + 24 - 1$

Notice that $28 - 1 = 27$, $26 - 1 = 25$, and $24 - 1 = 23$.

Both sides are the same. True.

(b) $16 + (1 - 16) = (16 - 16) + 1$

On both sides, we add a 16 and then subtract a 16 (cross out both). Both sides are equal to 1. True.

(c) $11 - (5 + 6) = 11 - 5 + 6$

On the left hand side, we subtract 5 and 6 from 11. The left hand side is equal to 0. On the right hand side we subtract 5 from 11 and then add 6. The right hand side is equal to 12. False.

(d) $5 + 15 + 25 - 20 - 10 = 5 + (15 - 10) + (25 - 20)$

On both sides, we add 5, 15, and 25 and then subtract 10 and 20. True.

Exercise 9

Lila has to place parentheses to obtain the smallest possible result.

$$43 \quad - \quad 3 \quad + \quad 15 \quad - \quad 7$$

Solution 9

We want to subtract the largest number possible:

$$43 - (3 + 15) - 7 = 43 - 18 - 7 = 18$$

Exercise 10

Dina has to place parentheses to obtain the result shown:

$$49 \quad - \quad 8 \quad - \quad 41 \quad + \quad 37 \quad = \quad 45$$

Solution 10

Dina notices that $37 + 8 - 41 = 4$ and that by subtracting 4 from 49 she obtains the required result. She places parentheses so that the numbers that add up to 4 are grouped together.

$$49 - (8 - 41 + 37) = 45$$

Exercise 11

Compute the result:

$$2 + 4 + 6 + 8 + 10 - 1 - 3 - 5 - 7 - 9 =$$

Solution 11

$$2 + 4 + 6 + 8 + 10 - 1 - 3 - 5 - 7 - 9 = 5$$

278

Notice that:

$$10 - 9 = 1$$
$$8 - 7 = 1$$
$$6 - 5 = 1$$
$$4 - 3 = 1$$
$$2 - 1 = 1$$

Exercise 12

Dina has to place parentheses in the following expression so that the computation is correct:

Solution 12

$$(40 + 8) \div 8 = 6$$

Exercise 13

Lila has to place parentheses in the following expression so that the computation is correct:

Solution 13

$$(9 + 8) \times 3 = 51$$

Exercise 14

Place parentheses in the following expression so that the computation is correct:

Solution 14

$$(8 + 10) \times 2 = 36$$

Exercise 15

Place parentheses in the following expression so that the computation is correct:

Solution 15

$$(4 + 5) \times (9 - 3) = 9 \times 6 = 54$$

Exercise 16

Without calculating, can you tell whether each line is true or false?

Solution 16

$10 - 1 + 11 - 1 = 10 + 11 - 2$	True
$8 - 3 = -3 + 8$	True
$8 - 3 = 3 - 8$	False
$5 + 3 - 3 = 8 + 5 - 8$	True
$11 - 12 + 12 - 1 = 11 - 1$	True

Exercise 17

True or false?

Solution 17

Pair the identical numbers that are both added and subtracted. Both sides are equal to zero.

$$2 - 2 + 3 - 3 + 4 - 4 + 5 - 5 = -2 + 2 - 3 + 3 - 4 + 4 - 4 + 5$$

True.

Exercise 18

True or false?

Solution 18

On both sides, we see pairs of numbers that are add-subtract pairs. Both sides are equal to zero.

$$5 + 6 + 7 + 8 + 9 - 5 - 6 - 7 - 8 - 9 = 5 + 6 + 7 + 8 + 9 - (5 + 6 + 7 + 8 + 9)$$

SOLUTIONS TO PRACTICE THREE

Exercise 1

$$\{2,\ 4,\ 6,\ 8,\ 10,\ \cdots,20\}$$

(a) How many numbers are written out?

(b) How many numbers have been replaced by dots?

(c) How many numbers are there in total (written and unwritten)?

Solution 1

The complete list is $\{2,\ 4,\ 6,\ 8,\ 10,\ 12,\ 14,\ 16,\ 18,\ 20\}$.

(a) 6 numbers are written out in the shortened list.

(b) 4 numbers have been replaced by dots.

(c) The list contains 10 numbers in total.

Exercise 2

$$\{2,\ 4,\ 6,\ 8,\ 10,\ \cdots,40\}$$

(a) How many numbers are written out?

(b) How many numbers have been replaced by dots?

(c) How many numbers are there in total (written and unwritten)?

Solution 2

This problem has more numbers than the previous problem to show students how important it is to make the transition from brute force to strategic reasoning.

Brute force solution:

The complete list is:

$$\{2,\ 4,\ 6,\ 8,\ 10,\ 12,\ 14,\ 16,\ 18,\ 20,\ 22,\ 24,\ 26,\ 28,\ 30,\ 32,\ 34,\ 36,\ 38,\ 40\}$$

(a) 6 numbers are written out.

(b) 14 numbers have been replaced by dots.

(c) The list contains 20 numbers in total.

Strategic solution:

Notice that the list is made up of consecutive even numbers. Figure out how many even numbers there are from 1 to 40. Half of them (20) are even. This is how many numbers there are in the list that is written out completely. In the abbreviated list, only 6 numbers are written out. $20 - 6 = 14$ numbers have been replaced by dots.

Exercise 3

Can you find out how many numbers have not been written out?

$$\{2,\ 4,\ 6,\ 8,\ 10,\ \ldots, 400\}$$

Solution 3

There must be 200 numbers in total. 6 numbers have been written out. The remaining $200 - 6 = 194$ numbers have been replaced by dots.

Exercise 4

Dina has to answer some questions about the following expression:

$$1 + 2 + 3 + \cdots + 40$$

1. How many numbers are added together?

2. How many of these numbers are odd?

3. How many + operators are there in total?

4. Will the result be even or odd? (Answer without calculating the result.)

Solution 4

1. 40 numbers have been added together.

2. Half of them (20 numbers) are odd.

3. Operators only occur between numbers (remember what you learned in "Practice Counting"). There are 39 plus signs.

4. Since there are 20 odd terms, they can be paired to form even sums. All the even numbers will add up to an even number. The final result will be even.

Solution 5
3 squares are the same as (equivalent to) one triangle.

$$\square + \square + \square + \square = \triangle + \square$$

Solution 6
100 squares are the same as (equivalent to) one star.

$$\underbrace{\square + \square + \cdots + \square}_{\text{101 squares}} = \bigstar + \square$$

Solution 7
22 squares are the same as (equivalent to) one circle.

$$\underbrace{\square + \square + \cdots + \square}_{\text{66 squares}} = \bigcirc + \bigcirc + \bigcirc$$

283

Exercise 8

Dina has to count the number of operations in the following sum:

$$1 + 2 + 3 + 4 + 5 + 6 + 7 + 8 + 9 + 10$$

Solution 8

There are 10 numbers to add. Therefore, 9 additions (operations) are necessary.

Exercise 9

Lila has to count the number of operations in the following sum:

$$1 + 2 + 3 + 4 + 5 + \cdots + 100$$

If Lila knows what Dina found in the previous problem, can she find an answer without writing out all the numbers?

Solution 9

There are 100 numbers to add. Therefore, 99 additions (operations) are necessary.

$$1 + 2 + 3 + 4 + 5 + \cdots + 100$$

Exercise 10

Compute efficiently:

$$2 + 4 + 6 + 8 + 10 + \cdots + 50 - 1 - 3 - 5 - 7 - 9 - \cdots - 49 =$$

Solution 10

Pair the terms in a different way:

$$2 - 1 + 4 - 3 + 6 - 5 + 8 - 7 + 10 - 9 + \cdots + 50 - 49 =$$

and notice how each pair is equal to 1.

Since there are 50 numbers in total, there must be 25 pairs.

The sum of the numbers in each pair is equal to 1, so the total is 25.

Exercise 11

How many numbers are there in the list?

$$\{11,\ 22,\ \cdots,99,\ 111,\ 222,\ \cdots,999\}$$

Solution 11

There are nine 2-digit numbers and nine 3-digit numbers.
There are $9 + 9 = 18$ numbers in total.

Exercise 12

How many numbers are there in the list?

$$\{1,\ 11,\ 111,\ 1111,\ \cdots,111111111111111\}$$

Solution 12

Each number has one more digit than the preceding one. Since the last number has 15 digits, there are 15 numbers in the list.

Exercise 13

How many numbers in the list are even?

$$\{0,\ 1,\ 2,\ 3,\ \cdots,100\}$$

Solution 13

From 1 to 100 there are 50 even numbers. Since zero is also even, the total number of even numbers in the list is 51.

Exercise 14

How many towers are there?

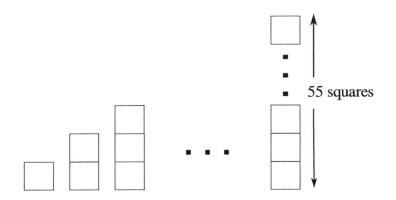

Solution 14

Each tower is one block higher than the previous tower. There are 55 towers in total.

Exercise 15

How many circles are filled?

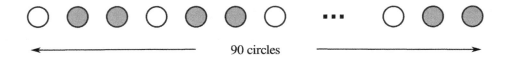

Solution 15

The circles can be placed in groups of three, with one white and two shaded circles in each group.

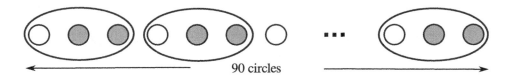

There are 30 groups of three in 90. In each group of three, two circles are shaded. The total number of shaded circles is 60.

Exercise 16

How many digits have been used to write the list:

$$\{1, \ 22, \ 333, \ \cdots, 7777777, \ \cdots, 333, \ 22, \ 1\}$$

Solution 16

The number of digits used is equal to the sum:

$$1 + 2 + 3 + 4 + 5 + 6 + 7 + 6 + 5 + 4 + 3 + 2 + 1 =$$

Since:

$$1 + 2 + 3 + 4 + 5 + 6 = 21$$

We have this sum twice plus a 7. The total is:

$$21 + 21 + 7 = 49$$

49 digits have been used.

Note: The problem did not ask for the number of *different (distinct)* digits. If that had been the case, the answer would have been 7.

Exercise 17

Lila and Dina are playing a game of dots. They use a ruled sheet of paper. Dina starts by drawing a dot on the first line. Lila draws two dots on the second line. Dina draws three dots on the third line. They continue on until one of them draws 47 dots on a line. Which line is it? Who draws the 47 dots, Lila or Dina?

Solution 17

The number of dots on each line is the same as the line number. Since Lila and Dina take turns drawing the dots, Lila will draw the dots on all the even numbered lines and Dina will draw the dots on all the odd numbered lines. Therefore, Dina will draw 47 dots on the 47th line.

Solution 18

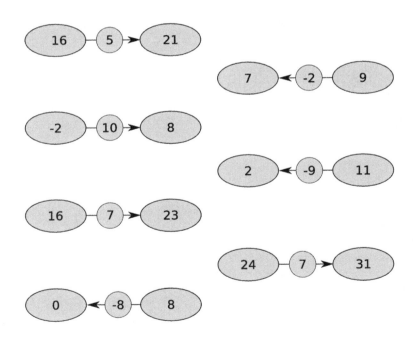

SOLUTIONS TO PRACTICE FOUR

Exercise 1

Which number is hiding behind the square?

Solution 1

$20 + 20 = 40$

Exercise 2

Which number is hiding behind the circle?

Solution 2

$11 + 11 = 22$

Exercise 3

Fill in the circles with appropriate non-zero positive integer numbers:

Solution 3

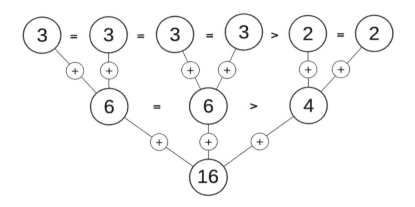

Exercise 4

Find K:

$$61 + 59 = K + K + K$$

Solution 4

$$\begin{aligned} 120 &= K + K + K \\ 120 &= 30 + 30 + 30 \\ K &= 30 \end{aligned}$$

Exercise 5

Place the same number in each empty circle ('/' means 'divide'). Which number must it be?

Solution 5

Exercise 6

Place in the circles numbers formed using the same digit:

Solution 6

Exercise 7

Place operators $(+, -, \times, \div)$ within the boxes. No parentheses are needed.

Solution 7

Exercise 8

In each of the following, which number does the letter represent?

$$A + A + A = 15$$

$$B + B + B + B - B = 15$$

$$C + C + C + C + C = 15$$

$$C + C + C + C + C + C - C = 15$$

$$D + D - D + D - D + D - D + D - D + D - D = 15$$

Solution 8

Use the fact that repeated addition is multiplication. Also, adding and subtracting the same number from a number leaves it unchanged. Cross-out add/subtract pairs of identical numbers.

$$
\begin{aligned}
A &= 5 \\
B &= 5 \\
C &= 3 \\
C &= 5 \\
D &= 15
\end{aligned}
$$

292

Exercise 9

The triangle, the circle, and the square have different weights. Which of the following would balance the scale on the right?

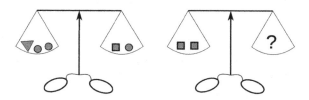

Solution 9

The balance on the left contains one triangle and two circles in one pan, and one square and one circle in the other pan. Since the two pans are perfectly balanced, one square must weigh as much as one triangle and one circle. Therefore, two squares will have the same weight as two triangles and two circles. The correct answer is (D).

Exercise 10

Amira is out shopping for Hallowe'en treats. She wants to buy some Real Fruit Eyeballs and some Fruit Jelly Smellies. Three packs of Eyeballs and two packs of Smellies cost 9 dollars. Two packs of Eyeballs and three packs of Smellies cost 11 dollars. Amira has 12 dollars. Does she have enough money to buy three packs of Eyeballs and three packs of Smellies?

Solution 10

Five packs of Eyeballs and five packs of Smellies cost $9 + 11 = 20$ dollars. Therefore, one pack of Eyeballs and one pack of Smellies cost 4 dollars and two packs of Eyeballs and two packs of Smellies cost 8 dollars. Since we know that two packs of Eyeballs and three packs of Smellies cost 11 dollars, one pack of Smellies must cost 3 dollars. This means that a pack of Eyeballs costs one dollar. Three packs of Eyeballs and three packs of Smellies cost exactly 12 dollars. Amira can make her purchase!

Exercise 11

Dina and Lila have traveled to their grandparents' home a total of 30 times. Of these, Lila traveled on 6 fewer occasions than Dina. How many times did each of them visit their grandparents?

Solution 11

The sum of their visits is 30 and the difference is 6. Subtract the difference from the sum:

$$20 - 6 = 24$$

The half of 24 is the smaller number. Lila traveled 12 times and Dina traveled 18 times.

Exercise 12

A number is smaller than another by 45. The sum of the two numbers is 99. Find the two numbers.

Solution 12

The sum of the numbers is 99 and their difference is 45. Subtract the difference from the sum:

$$99 - 45 = 54$$

The half of 54 is the smaller number. If you do not know how to divide simply make equal portions in the way you know best. I would do:

$$50 = 25 + 25$$
$$4 = 2 + 2$$

The half is 27. One of the numbers is 27 and the other is $45 + 27 = 72$.

Exercise 13

Arbax has given Lynda 12 of his treats. Now, Lynda has 5 treats more than Arbax. How many treats more than Lynda did Arbax have?

Solution 13

If Lynda has now 5 treats more than Arbax, this means that by the time Arbax had given her only 7 treats, they had equal numbers of treats. Therefore, Arbax had 14 more treats than Lynda.

Exercise 14

Amira has lost 14 of her action figures. Dina has lost only 10 of her action figures. Lila tells them: 'We used to have 104 action figures together. Now that you lost some of yours, I have as many action figures as both of you.' How many action figures does Lila have now?

Solution 14

After losing $10 + 14 = 24$ action figures, the total has become $104 - 24 = 80$. Since Lila has as much as both the other girls, she must have half of the total, which is 40. Notice that we cannot find out how many action figures Dina or Amira have, only that their action figures total to 40.

Exercise 15

Lila, Dina, and Amira are picking flowers to decorate the living room. Lila picks 5 more flowers than Dina. Amira picks 3 flowers less than Dina. The whole bunch is made up of 23 flowers. How many flowers did each of them pick?

Solution 15

Use comparison to establish that Amira picks the least amount of flowers. Make a box to represent this amount:

Amira

Now, make boxes to represent the amounts the other two girls picked:

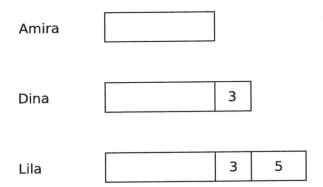

Compare to the total:

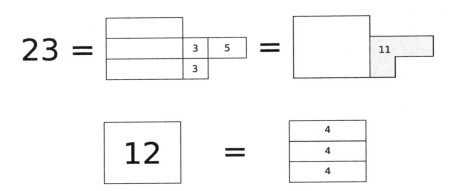

Amira has picked 4 flowers, Dina has picked $4 + 3 = 7$, and Lila has picked $4 + 3 + 5 = 12$ flowers.

SOLUTIONS TO MISCELLANEOUS PRACTICE

Exercise 1

Which number does the circle represent?

$$\bigcirc + \bigcirc = \bigstar$$

$$\bigstar + \bigstar = 136$$

Solution 1

Each circle is the same as half a star. Each star represents half of 136. Therefore, the star represents 68 and the circle represents 34.

Exercise 2

Replace the question mark with a number so that the operations are correct:

$$\bigcirc + \bigstar = 16$$

$$\bigcirc - \bigstar = 4$$

$$\bigcirc + \bigcirc = ?$$

Solution 2

In the first operation, the star is added to the circle and in the second one, the star is subtracted from the circle. Therefore, $16 + 4$ equals two circles, plus a star, minus a star. Adding and subtracting the same amount does not change anything. Cross-out the add-subtract pair of the same symbol. Two circles equal 20.

Exercise 3

If two positive integers have an odd difference, is their sum:

(A) always even?

(B) always odd?

(C) sometimes even and sometimes odd?

Solution 3

If their difference is odd, then the numbers must have different parity. One must be even, while the other must be odd. Therefore, their sum must also be odd. The answer is (B).

Exercise 4

There are 100 plus signs in the following operation. What is the result of the additions?

$$1+1+1+\cdots+1= ?$$

Solution 4

If there are 100 operators, then there must be 101 terms, each equal to 1. The sum is equal to 101.

Solution 5

Roman numerals time! Perform the following operations and write the answer in both Roman and Arabic numerals:

(a) $I + I = II$ $(1 + 1 = 2)$

(b) $II + II = IV$ $(2 + 2 = 4)$

(c) $III + III = VI$ $(3 + 3 = 6)$

(d) $IV + IV = VIII$ $(4 + 4 = 8)$

(e) $V + V = X$ $(5 + 5 = 10)$

(f) $XX + XX = XL$ $(20 + 20 = 40)$

(g) $XXX + XXX = LX$ $(30 + 30 = 60)$

(h) $LX + LX = CXX$ $(60 + 60 = 120)$

    (i) $L + L = C$ $(50 + 50 = 100)$

    (j) $C + C = CC$ $(100 + 100 = 200)$

    (k) $CC + CC = CD$ $(200 + 200 = 400)$

    (l) $CCC + CCC = DC$ $(300 + 300 = 600)$

  (m) $CD + CD = DCCC$ $(400 + 400 = 800)$

  (n) $CD + D = CM$ $(400 + 500 = 900)$

Exercise 6

Each group of four numbers in the sequence follow the same pattern. Which numbers belong in the empty circles?

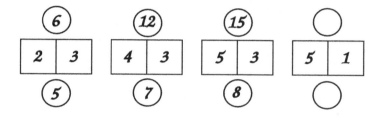

Solution 6

The circle on top contains the product of the numbers in the squares. The circle on the bottom contains the sum of the numbers in the squares:

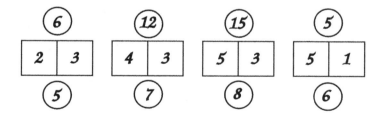

Exercise 7

In the figure, shaded circles of the same color hide the same number. Which number does the white circle hide?

Solution 7

The sum of the six white circles is 12. Therefore, each white circle must hide a 2. The intermediate values can be determined but are not necessary in deriving the answer.

Exercise 8

Compute:
$$4 + 3 - 4 + 5 - 6 + 6 - 7 + 8 - 9 =$$

Solution 8

Notice how the terms can be grouped in pairs that are equivalent to subtracting 1. Each of the operations $3 - 4$, $5 - 6$, $6 - 7$, and $8 - 9$ subtracts a 1. We have to subtract 1 four times from the first term. The result is zero.

Exercise 9

Dina had 30 party balloons. During the party, 11 balloons popped. Afterwards, Lila gave Dina as many balloons as had popped. How many party balloons did Dina have then?

Solution 9

In this problem, we subtract a number and then add it back. These operations do not change the initial number. Dina had 30 balloons after Lila gave her as many balloons as had popped.

Exercise 10

The table in the figure has 10 rows and 11 columns. How many grey cells are there in the table? How many grey cells should we color white in order to have an equal number of white and grey cells?

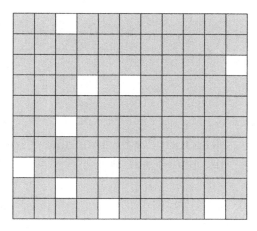

Solution 10

Notice that there are 10 white squares. The table has 10 rows and 11 columns, therefore it must have 110 cells. Subtract $110 - 10 = 100$ to obtain the number of grey cells.

There are 90 more grey cells than white cells. In order to have an equal number of white and grey cells, we need to paint 45 cells white.

Exercise 11

In the lists below, some digits, not necessarily identical, have been hidden by a ♠. In each list, there are two different 3-digit numbers. Find the lists in which it is possible to tell which number is smaller.

(A) 9♠7, ♠87

(B) 33♠, 31♠

(C) 4♠3, 5♠♠

(D) 9♠♠, 9♠0

Solution 11

Only in lists (B) and (C) it is possible to tell which number is smaller without knowing which digits are hidden by the ♠s.

Exercise 12

Dina has 20 magnets and Lila has 11 magnets. How many magnets could Amira have if she has fewer than Dina and more than Lila? Check all that apply.

(A) 9

(B) 10

(C) 13

(D) 19

Solution 12

The number of magnets Lila has must be larger than 11 and smaller than 20. Possible numbers are: 12, 13, 14, 15, 16, 17, 18, and 19. Answer choices (C) and (D) are in this list.

Exercise 13

How many 3-digit numbers have a digit sum of 2?

Solution 13

Three numbers: 110, 101, and 200.

Exercise 14

A 100-digit number has a digit sum of one. How many digits 0 does the number have?

Solution 14

Since zero cannot be the first digit of a number, the number must start with a digit of 1 or greater. Because the digit sum must be 1, then the number must start with a 1 and all the remaining digits must be zero. The 1 is followed by 99 zeros.

Exercise 15

Dina has written a 3 digit number on a piece of paper. Lila must guess which number it is. Dina gives Lila a hint: "One of the digits is 5. The number does not change if you move the last digit in front of the first." Which number is it?

Solution 15

By moving the last digit to the leftmost position, the middle digit becomes the last. For the number to remain unchanged, therefore, the middle digit must equal the last digit. Also, the leftmost digit becomes the middle digit and, therefore, the leftmost digit must equal the middle digit. As a result, all the digits must be identical. The number must be 555.

Solution 16

Compute:

(a) $10 - 9 + 8 - 7 + 6 - 5 + 4 - 3 + 2 - 1 + 0 = 1 + 1 + 1 + 1 + 1 = 5$

(b) $100 + 99 - 99 + 98 - 98 + 97 - 97 + 96 - 96 + 95 - 95 + 94 - 94 + 93 - 93 + 92 - 92 + 91 - 91 + 1 = 101$

(c) $9 - 2 + 3 - 4 + 5 = 9 + 1 + 1 = 11$

Exercise 17

Lila has 5 more toys than Amira. If Lila gives Amira 7 toys, how many more toys than Lila will Amira have?

Solution 17

Amira would have 9 more toys than Lila:

Exercise 18

Find the positive integers that are hidden by symbols. Different symbols correspond to different integers. How many different solutions can be found?

$$\diamondsuit + \heartsuit + \heartsuit = 7$$

Solution 18

3 solutions can be found:

$$1 + 3 + 3 = 7 \qquad 3 + 2 + 2 = 7 \qquad 5 + 1 + 1 = 7$$

Exercise 19

In the following sequence of consecutive odd numbers, how many numbers have been replaced by dots?

$$15, \ 17, \ \cdots, 25$$

Solution 19

There are 3 numbers missing: 19, 21, and 23.

Exercise 20

How many numbers between 200 and 240 can be written using only the digits 2 and 3?

Solution 20

Four numbers: 222, 223, 232, and 233.

Exercise 21

A machine crunches numbers and outputs the result. The figure shows a set of numbers entering the machine and the results of the crunches coming out of the machine. If the machine is given the number 207 to crunch, what will it output?

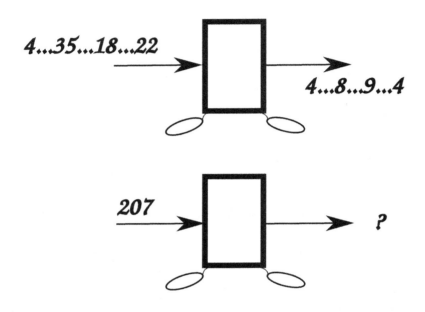

Solution 21

Notice that the machine outputs the digit sum of the numbers that it crunches. Therefore, if the input is 207, the machine will output 9.

Exercise 22

Another machine mashes numbers and outputs the result. The figure shows a set of numbers entering the machine and the results of the mashes coming out of the machine. If the machine is given the number 207 to mash, what will it output?

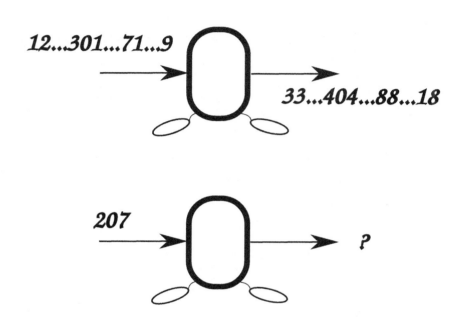

Solution 22

Notice that the machine reverses the digits of the number and then adds the new number to the initial one:

$$
\begin{aligned}
12 + 21 &= 33 \\
301 + 103 &= 404 \\
71 + 17 &= 88 \\
9 + 9 &= 18
\end{aligned}
$$

Therefore, if the input is 207, the machine will output $207 + 702 = 909$.

Math Challenges for Gifted Students

Practice Tests in Math Kangaroo Style for Students in Grades $1-2$

Practice Tests in Math Kangaroo Style for Students in Grades $3-4$

Practice Tests in Math Kangaroo Style for Students in Grades $5-6$

Weekly Math Club Materials for Students in Grades $1-2$

Competitive Mathematics Series for Gifted Students

Practice Counting (ages 7 to 9)

Practice Logic and Observation (ages 7 to 9)

Practice Arithmetic (ages 7 to 9)

Practice Operations (ages 7 to 9)

Practice Word Problems (ages 9 to 11)

Practice Combinatorics (ages 9 to 11)

Practice Arithmetic(ages 9 to 11)

Practice Operations (ages 9 to 11)

Practice Word Problems (ages 11 to 13)

Practice Combinatorics and Probability (ages 11 to 13)

Practice Arithmetic and Number Theory (ages 11 to 13)

Practice Algebra and Operations (ages 11 to 13)

Practice Geometry (ages 11 to 13)

Self-help:

Parents' Guide to Competitive Mathematics

Coming Soon:

Weekly Math Club Materials for Students in Grades $3 - 4$
Weekly Math Club Materials for Students in Grades $5 - 6$

Practice Word Problems (ages 12 to 16)
Practice Algebra and Operations (ages 12 to 16)
Practice Geometry (ages 12 to 16)
Practice Number Theory (ages 12 to 16)
Practice Combinatorics and Probability (ages 12 to 16)

This is a series of practice books. With the exception of a few reminders, there are no theoretical explanations. Online problem solving lessons are available at www.goodsofthemind.com. If you found this booklet useful, you will enjoy the live problem solving lessons.

For additional problem solving material, look up our series *Math Challenges for Gifted Students* which includes practice tests in Math Kangaroo style and weekly math club materials.

Made in the USA
Coppell, TX
28 March 2020